Hendrik Kersten

Lightning meets Mass Spectrometry

Hendrik Kersten

Lightning meets Mass Spectrometry

Development of a windowless spark discharge and laser ionization source operating at atmospheric pressure

Südwestdeutscher Verlag für Hochschulschriften

Impressum/Imprint (nur für Deutschland/only for Germany)
Bibliografische Information der Deutschen Nationalbibliothek: Die Deutsche Nationalbibliothek verzeichnet diese Publikation in der Deutschen Nationalbibliografie; detaillierte bibliografische Daten sind im Internet über http://dnb.d-nb.de abrufbar.
Alle in diesem Buch genannten Marken und Produktnamen unterliegen warenzeichen-, marken- oder patentrechtlichem Schutz bzw. sind Warenzeichen oder eingetragene Warenzeichen der jeweiligen Inhaber. Die Wiedergabe von Marken, Produktnamen, Gebrauchsnamen, Handelsnamen, Warenbezeichnungen u.s.w. in diesem Werk berechtigt auch ohne besondere Kennzeichnung nicht zu der Annahme, dass solche Namen im Sinne der Warenzeichen- und Markenschutzgesetzgebung als frei zu betrachten wären und daher von jedermann benutzt werden dürften.

Coverbild: www.ingimage.com

Verlag: Südwestdeutscher Verlag für Hochschulschriften GmbH & Co. KG
Dudweiler Landstr. 99, 66123 Saarbrücken, Deutschland
Telefon +49 681 37 20 271-1, Telefax +49 681 37 20 271-0
Email: info@svh-verlag.de

Zugl.: Wuppertal, Diss., 2011

Herstellung in Deutschland:
Schaltungsdienst Lange o.H.G., Berlin
Books on Demand GmbH, Norderstedt
Reha GmbH, Saarbrücken
Amazon Distribution GmbH, Leipzig
ISBN: 978-3-8381-2799-6

Imprint (only for USA, GB)
Bibliographic information published by the Deutsche Nationalbibliothek: The Deutsche Nationalbibliothek lists this publication in the Deutsche Nationalbibliografie; detailed bibliographic data are available in the Internet at http://dnb.d-nb.de.
Any brand names and product names mentioned in this book are subject to trademark, brand or patent protection and are trademarks or registered trademarks of their respective holders. The use of brand names, product names, common names, trade names, product descriptions etc. even without a particular marking in this works is in no way to be construed to mean that such names may be regarded as unrestricted in respect of trademark and brand protection legislation and could thus be used by anyone.

Cover image: www.ingimage.com

Publisher: Südwestdeutscher Verlag für Hochschulschriften GmbH & Co. KG
Dudweiler Landstr. 99, 66123 Saarbrücken, Germany
Phone +49 681 37 20 271-1, Fax +49 681 37 20 271-0
Email: info@svh-verlag.de

Printed in the U.S.A.
Printed in the U.K. by (see last page)
ISBN: 978-3-8381-2799-6

Copyright © 2011 by the author and Südwestdeutscher Verlag für Hochschulschriften GmbH & Co. KG and licensors
All rights reserved. Saarbrücken 2011

Table of Contents

Table of Contents i

Notation and Terminology vi

1 Introduction 1

   1.1 Atmospheric Chemistry 1

      1.1.1 Laboratory studies 2

   1.2 Mass Spectrometry 3

      1.2.1 Atmospheric Pressure Ionization (API) 4

         1.2.1.1 Atmospheric Pressure Laser Ionization (APLI) 4

         1.2.1.2 Atmospheric Pressure Photo Ionization (APPI) 5

         1.2.1.3 Negative Ion Formation (NIF) 6

         1.2.1.4 Ion Transformation Processes (ITP) 7

2 Goals 11

3 Experimental 12

   3.1 Mass Spectrometer 12

      a) Storing mass range capability. 14

      b) Mass discrimination 15

      c) Mass analysis 15

      d) Mass resolution 15

      e) Mass accuracy 16

      f) Switching positive-negative modus 16

      g) Resonant excitation 16

      h) Ion isolation 17

      i) CID 17

      j) Ms^n experiments 17

      k) Duty cycle 18

Table of Contents

  l) Chromatogram mode 18

  m) Software 18

 3.1.1 Laser Systems 18

 3.1.2 Common API Source 20

 3.1.3 Novel Laminar-Flow Ion Source (LFIS) 21

 3.1.4 Novel APPI Setup 23

  3.1.4.1 Setup for Characterization of Transfer Capillaries 24

  3.1.4.2 Characterization of the Discharge Lamp 25

  3.1.3 Setup for Neutral Radical Induced ITP Studies [82] 26

3.3 Photoreactor 27

 3.3.1 Procedure of Atmospheric Degradation Studies 27

 3.3.2 FT-IR-setup 28

 3.3.3 MS Sampling Unit 28

 3.3.4 MS Ionization Unit 29

3.4 Chemicals 30

3.5 Computational Investigations 30

4 Results and Discussion 31

 4.1 Common API sources 31

  4.1.1 Distribution of Ion Acceptance (DIA) Studies 31

  4.1.2 Fluid Dynamical Behavior 33

  4.1.3 H_2O and O_2 Background Concentrations [82] 35

 4.2 Development of a Novel API Approach 37

  4.2.1 LFIS - Preliminary Experiments 37

   a) Rough determination of the flow characteristics 37

   b) Ion transmission efficiencies 38

   c) Different behavior of quartz and metal tubes 40

   d) Impact of the laser frequency in coaxial configuration 40

Table of Contents

- e) Suggestive estimate of coaxial sensitivity 41
- 4.2.2 LFIS - Realization 42
  - 4.2.2.1 LFIS - Fluid Dynamical Simulations 43
    - a) Flow characteristic 43
    - b) Diffusion along the flow propagation 45
  - 4.2.2.2 LFIS – APLI 46
    - a) Choice of laser system 46
    - b) Laser beam expansion. 47
    - c) Interaction laser radiation → metal surface 51
  - 4.2.2.3 LFIS – APPI 52
    - a) Implementation 52
    - b) Transit time. 53
- 4.2.3 Development of APPI on Transfer Capillaries 54
  - 4.2.3.1 Characterization of Transfer Capillaries 55
    - a) Comparison of original and home–made capillary 55
    - b) Adaptability of fluid dynamic equations - laminar or turbulent 57
    - c) Critical and static pressure, velocity distribution, and transit times 58
    - d) Upstream pressure variation 60
  - 4.2.3.2 First APPI on Capillary Approach 62
  - 4.2.3.3 Development of Miniature VUV Spark Discharge Lamps 63
    - a) In general 63
    - b) High voltage-power supplies 63
    - c) APPI with or without window 66
    - d) Balanced pressure separation 68
    - e) Lamp design 1 69
    - f) Lamp design 2 70
    - g) Lamp design 3 71

Table of Contents

 h) Operating stability tests 72

 i) Determination of lower detection limits (LODs) 73

 j) Experimental sparking characteristics of design 3 with the DD20_10 C-Lader .. 74

 k) Theoretical considerations on the spark characteristics 76

 l) Optical emission spectroscopy (OES) 81

 m) VUV emission efficiency in comparison to the commercially available APPI lamp 86

 n) VUV emission spectroscopy below 105 nm 86

 4.2.3.4 Impact of Different Ionization Positions on MS Spectra 87

  a) Impact on negative ion mode 88

  b) Impact on positive ion mode 89

4.3 Ion Transformation Processes (ITP) 91

 4.3.1 Unintended Collision Induced ITP 91

 4.3.2 Neutral Radical Induced ITP (NRITP) [82] 93

  4.3.2.1 Evidence for Ion-Neutral Radical Chemistry 95

  4.3.2.2 Oxidation of the Pyrene Radical Cation - Feasible Pathways 96

   a) Oxidation via direct addition of $O(^3P)$ and OH 98

   b) Oxidation via the pyrenyl cation $[M-H]^+$ 100

  4.3.2.3 Oxidation of the Pyrene Radical Cation - Kinetic Investigations 102

   a) Impact of O_2, $O(^3P)$, and O_3 on the ion distribution 103

   b) Impact of H_2O, OH, and H on the ion distribution 104

   c) Impact of Cl, ClO, and ClOO on the ion distribution 105

  4.3.2.4 Consequences for Degradation Studies with APPI-MS 106

 4.3.3 ITP via Chemical Ionization 108

  4.3.3.1 APPI/APLI-Positive Ion Chemical Ionization (PICI) 109

  4.3.3.2 APPI/APLI-Negative Ion Chemical Ionization (NICI) 111

4.4 Degradation Studies 114

    4.4.1 Features and limitations of the MS setup 114

    4.4.2 Exemplary Degradation Study 115

        4.4.2.1 Blank test without p-xylene present in the reactor 115

            a) Signals of protonated water clusters 115

            b) Signals of NO_x, HNO_x and $HNO_x \cdot NO_x$ 115

            c) Signals of O_x 117

            d) Signals of CH_3ONO and its degradation products 118

        4.4.2.2 Degradation Study with p-xylene 119

            a) Initial step I: H-atom abstraction from methyl group 120

            b) Initial step II: OH-addition to the aromatic ring 122

5 Summary and Conclusion 126

        a) Investigations on the commercially available API source 126

        b) Development of a laminar flow ion source with a laminar sampling unit 126

        c) Novel APPI approach with home-built miniature spark discharge lamps 127

        d) Ion transformation processes 129

        e) Exemplary degradation study of p-xylene 130

6 Indexes ix

    6.1 List of Figures ix

    6.2 List of Tables xi

    6.3 References xii

Notation and Terminology

List of Acronyms

[F]	neutral fragment	CI	chemical ionization
[F]⁻	anionic fragment	CID	collision induced dissociation
[F]⁺	cationic fragment	DA	dopant assisted
[M]	neutral molecule	DC	direct current
[M]⁻	molecular anion	DFT	Density Functional Theory
[M]⁻*	excited molecular anion	DIA	distribution of ion acceptance
[M]⁺	molecular cation	DPSS	diode pumped solid state laser
[M+15/16/17]⁺	ion with 15/16/17 masses off the molecular ion mass	e^-	electron
[M+Cl]⁺	cation with an added Cl-atom	e.g.	for example
[M+H]⁺	quasi-molecular ion	EC	electron capture
[M+O]⁺	cation with an added O-atom	ECD	electron capture dissociation
[MH]	neutral molecule with an added H-atom	EI	electron impact ionization
		eq	equation
[M-H]⁻	deprotonated anion	ESI	electrospray ionization
[M-H]⁺	cation with an abstracted H-atom	et al.	and co-workers
[M-H+Cl]⁺	cation with an abstracted H-atom and an added Cl-atom	etc.	and so on
[M-H+O]⁺	cation with an abstracted H-atom and an added O-atom	FT-IR	Fourier transform infrared spectroscopy
[MP]	neutral cluster of M and P	FWHM	full-width-half-maximum
[MPH]	neutral cluster of M and P with an added H-atom	GC	gas chromatography
		HPLC	high pressure liquid chromatography
[NR]	neutral radical	HV	high voltage
[P]	neutral product molecule	hν	photon
[P]⁺	cation product molecule	i.e.	that means
[P+H]⁺	quasi-molecular ion of product	IC	ion chromatography
		ICC	ion charge control
a.u.	arbitrary units	IR	infra red
AC	alternating current	ITP	ion transformation process
ACN	acetonitrile	LC	liquid chromatography
AE	appearance energy	LFIS	laminar flow in source
AH	aromatic hydrocarbon	LID	laser induced dissociation
APCI	atmospheric pressure chemical ionization	lit.	literature
		LMCO	lower mass cut-off
API	atmospheric pressure ionization	LOD	lower detection limit
APLI	atmospheric pressure laser ionization	m/z	mass-to-charge ratio
APPI	atmospheric pressure photo ionization	MCM	Master Chemical Mechanism
		MeONO	methylnitrite
cf.	compare	MPIS	multi-purpose ion source
CFD	computational fluid dynamic	MS	mass spectrometry

Ms^n

Notation and Terminology

1 Introduction

1.1 Atmospheric Chemistry

A truely fascinating characteristic of our planet is that it provides exactly the right conditions for living species. Temperature, pressure, the air composition, to name but a few, all seem to be perfectly tailored. Considering the complexity and dynamics behind the coherence of the atmosphere, the solar radiation, the oceans, the weather, the flora, the fauna and us, the human beings, the robustness of this system appears to be quite incredible. For instance, the emission of volatile organic compounds (VOCs) from anthropogenic and non-anthropogenic sources to the atmosphere accounts for petagrams per year [1,2] but the tremendous oxidative chemistry within the tropospheric layer results in an effective degradation and removal. The responsible "washing agents" O_2, the trace gases NO_x and O_3, the OH radical and the solar radiation provide an elaborated "machinery" in which each compound undergoes its own individual photochemical degradation process [1,2] (cf. Figure 1, left). It all seems to be well balanced. However, key words like smog or ozone hole have rung the alarm bells and have revealed some kind of maladjustment. The need for a more profound understanding of atmospheric chemical processes has been obvious ever since. However, when considering the enormous diversity of VOCs and their individual chemical fates this challenge is more likely a puzzle with an overwhelming number of pieces. Nevertheless, atmospheric chemists all over the world have started to find those pieces and bring them together.

Since aromatic hydrocarbons (AH), predominantly benzene, toluene and the xylene isomers account for up to 44 % of the total VOC emission in cities and for up to 30 % of the total ozone formation in urban areas, much attention has been paid to this compound class [3,4]. The traffic, various industrial processes and the use of organic solvents constitute the main sources of aromatic compounds. The initial step in the atmospheric oxidation pathway of AHs is the reaction with OH radicals (cf. Figure 1, right). Two feasible degradation mechanisms are discussed in the literature: (i) The H-atom abstraction from alkyl substituent groups which accounts for approximately 10 % and (ii) the OH addition to the aromatic ring (OH-adduct) which accounts for the remainder [1]. The latter reacts in a subsequent step with molecular oxygen but in spite of intensive research efforts the ensuring products from this

1 Introduction

step are still poorly characterized [5]. Ring-opening is found to be a major channel and leads to formation of diverse saturated and unsaturated carbonyl compounds [6-8]. The interaction of the OH-adduct with O_2 also leads to formation of ring-retaining products. For benzene and toluene the ring-retaining products phenol and cresol isomers are formed with yields of up to 50 % [9,10] and 20 % [4], respectively. Nitro-phenols are also formed in significant yields [5]. The primary H-atom abstraction channel from the alkyl side groups also leads to the formation of ring-retaining products, e.g., in the case of toluene, benzaldehyde and benzylalcohol [5]. In addition there is growing evidence that the ring-retaining products play an important role with respect to secondary organic aerosol (SOA) formation [11]. These aerosols significantly impact on cloud formation processes and consequently impact the weather and the solar radiation balance. Furthermore negative health effects are related to SOA concentration increase.

It becomes clear that a complex coherence between the atmospheric chemistry and eventually life on earth is present. Elucidation of such mechanisms is only in its early stages and huge research efforts toward a better understanding are needed.

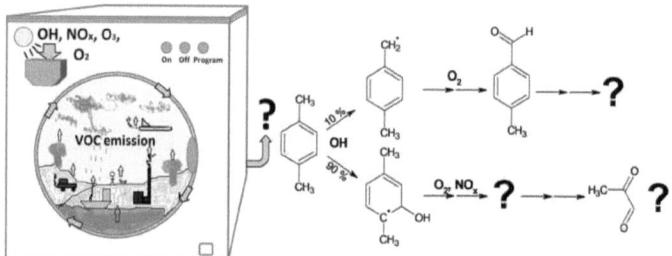

Figure 1: (left) Atmospheric "washing machine". (right) Excerpts of the photo oxidation pathway of p-xylene.

1.1.1 Laboratory studies

Investigations regarding atmospheric chemical processes are generally performed in the laboratory under well known and precisely controllable conditions. The obtained experimental data serve as input parameters for computational modeling, such as the Leeds Master Chemical Mechanism (MCM) [12,13], putting the puzzle pieces together. The validated mechanism will eventually allow for reliable predictions and trends, albeit, this ideal situation seems to lie far in the future. Current political decisions and regulations, however, are based on current and obviously imperfect models. Consequently, there is a demand for precise, highly accurate and reliable experimental data and their interpretation. The demand on sophisticated analytical instrumentation becomes thus self-evident.

The general setup for photochemical degradation experiments consists of a transparent chamber being exposed to sunlight, as realized in the EUPHORE project [14], or a chamber which is surrounded by lamps simulating the solar radiation [15,16]. The compound of interest and the "washing agents" or their photo labile precursors, respectively, are inserted into the chamber, which is then backfilled with synthetic air. The experiment is initiated by the exposure to visible/near-UV light. In the case of photochemical degradation of aromatic hydrocarbons miscellaneous analytical tools have been used, such as differential optical absorption spectroscopy (DOAS) [9,17,18], Fourier transform infrared spectroscopy (FT-IR) [19,20], ultraviolet (UV) absorption spectroscopy [21,22], gas chromatography (GC) [4], GC mass spectrometry (GC-MS) [4], derivatization and diverse sampling methods with subsequent GC-MS [6,8,23] analysis, ion chromatography (IC) [4] and in situ monitoring with proton transfer mass spectrometry (PTR-MS) [24].

1.2 Mass Spectrometry

In 1913 J.J. Thomson laid the foundation for mass spectrometry with his work on "rays of positive electricity", wherein he deduced the mass-to-charge ratio (m/z) of charged particles from their specific parabolic trajectories within electromagnetic fields [25]. A powerful analytical tool was born that with progression of the computer technology in the 1960s has meanwhile emerged into nearly every scientific research field which requires molecular measurements [26-28]. However, the use of mass spectrometry, besides PTR-MS [24], for in situ monitoring of photochemical degradation experiments with aromatic hydrocarbons is not yet widely spread.

The basic working principle of a mass spectrometer encompasses three sections: (i) Ionization, (ii) mass analysis, and (iii) ion detection. Each section, in particular the first two, has been subject to intensive research which has consequently led to a tremendous

1 Introduction

variety of instruments over the years [29-32]. A more detailed description of the mass analyzing principle of the quadrupole ion trap (QIT), as used in this work, will be given in the experimental section (cf. *3.1 Mass Spectrometer*).

1.2.1 Atmospheric Pressure Ionization (API)

One possible classification of the various existing ionization methods is based on the prevailing pressure. For a long time mass spectrometry was dominated by the classical low pressure ionization methods electron ionization (EI) ($p < 10^{-4}$ mbar) [29] and chemical ionization (CI) ($p \sim 1$ mbar) [33]. With the emerging demand for hyphenation with chromatographic stages, such as high performance liquid chromatography mass spectrometry (HPLC-MS) new ionization sources and methods operating at atmospheric pressure have been developed [34,35], for instance electrospray ionization (ESI) [34,35], atmospheric pressure chemical ionization (APCI) [34], atmospheric pressure photo ionization (APPI) [36] and atmospheric pressure laser ionization (APLI) [37]. API benefits from higher neutral analyte densities within the ionization region, which in general provides more sensitive methods than the classical low pressure techniques.

1.2.1.1 Atmospheric Pressure Laser Ionization (APLI)

This new API method was first introduced by Constapel et al. [37] in 2005 and became commercially available in 2008 [38]. The working principle is based on two photon-one-color (1+1) resonance enhanced multi photon ionization (REMPI) processes [29] (cf. Figure 2). Here, a molecule is ionized via a two step mechanism: (i) resonant electronic photo excitation and (ii) ionization through the absorption of a second photon within the lifetime of the primary excited state, provided that the sum energy of the two photons (taking into account intermediate relaxation processes) exceeds the ionization energy (E_i) of the molecule. The efficiency of this process is strongly dependent on the absorption cross section of the first step ($> 10^{-18}$ cm^2 molecule^{-1}), the lifetime of the primarily excited state ($> 10^{-9}$ s) and the photon density ($> 10^5$ W·cm^{-2}). Typical E_is of organic molecules range between 8 and 10 eV, which means that the radiation source should provide appreciable photon densities with wavelengths in the ultraviolet (UV) below 310 nm (> 4eV). In common APLI applications, a pulsed KrF* exciplex laser, radiating at 248 nm (5 eV) with power densities of 10^6 W·cm^{-2} is commonly used. Since the compound class of aromatic hydrocarbons exhibits fairly high

1 Introduction

absorption cross sections at this wavelength, APLI was introduced as a very selective and sensitive API method. With APLI, AHs are detectable two to three orders of magnitude more sensitive than with all other API methods. With GC-MS a limit of detection (LOD) in the attomol regime has been achieved [37,39,40]. Furthermore, APLI is compatible with direct infusion stages, liquid chromatography (LC) [37], as well as with atmospheric pressure gas chromatography [41]. Along this line, a rapid change from liquid chromatography (LC) to GC operation, change of ionization methods (ESI, APCI, APLI), as well as true multi-mode operation (ESI/APLI) was recently demonstrated with a novel multi-purpose ion source (MPIS) [41]. In addition, the amenable analyte range was successfully extended through derivatization with APLI-sensitive chromophoric molecular tags [42]. Despite the favorable performance of APLI, recent investigations concerning the distribution of ion acceptance (DIA) [43-47] and the fluid dynamic behavior [48] in typical API source enclosures revealed a considerable lack in overlap of the ionization volume and the neutral analyte distribution, and pinpointed deficiencies in ion transport into the MS.

1.2.1.2 Atmospheric Pressure Photo Ionization (APPI)

In 2000 Robb et al. [49] and Syage et al. [50] simultaneously introduced APPI as a new atmospheric pressure ionization method for LC-MS. Currently two APPI stages are commercially available fitting nearly every mass spectrometer that is delivered with an API interface [29,51]. For many applications this ionization technique has become the method of choice in particular for low to non-polar analytes. In contrast to the APLI approach, APPI makes use of single photon ionization processes with wavelengths in the vacuum ultraviolet (VUV) below 155 nm (> 8 eV), thus omitting the selective excitation step into a resonant state (cf. Figure 2). Consequently, a broader analyte range is amenable to this technique, and the photon densities as provided by simple gas discharge lamps are sufficient. A number of VUV radiation sources exist, however low pressure krypton radio frequency (Kr-RF) or krypton direct current (Kr-DC) discharge lamps have generally been accepted as the standard radiation source. Both produce two distinct VUV emissions centered at 123.6 nm (10.03 eV) and 116.5 nm (10.64 eV) in a 4:1 ratio. The reason for this preferred lamp choice is mainly argued with the larger than 10.6 eV E_is of the most common solvents used in liquid chromatography, whereas most organic molecules are amenable to photo ionization, i.e., their E_is are below 10.6 eV [51]. However, in contrast to the UV operating conditions in APLI, where most

1 Introduction

solvents are virtually transparent, in the VUV most solvents in addition to O_2 and H_2O exhibit absorption cross sections in the 10^{-17} cm^2 molecule^{-1} regime [52-58]. Thus, in typical LC-APPI measurements, the penetration depth of the radiation is less than 5 mm. The absorption spectra in this region are mostly structureless, indicating excited-state lifetimes far below the collision rate of about 10^9 s^{-1}. In other words, most of the available VUV energy applied in APPI is directly coupled into the matrix, leading to electronic excitations and/or extensive photolysis of matrix compounds and, hence, in formation of primary neutral radicals. Due to the low penetration depth poor direct photo ionization efficiencies of analytes have been reported [51]. This is why current APPI LC-MS applications are mainly performed with reactant liquids (~5 - 10 %), in analogy to the addition of reactant gases in classical chemical ionization, e.g. toluene, with E_is below the VUV photon energy rendering it amenable to photo ionization. In this way, photons are efficiently converted into charged species. Subsequent ion molecule interactions such as charge transfer or protonation lead to the formation of charged analytes. The latter method is known as dopant assisted (DA) APPI. As mentioned for APLI, APPI as well, as any classical API method suffers in the same manner from the disadvantages entailed by the typical API source enclosure designs.

1.2.1.3 Negative Ion Formation (NIF)

The first step in nearly all negative ion formation mechanisms is the generation of free electrons by photo electron emission from metal surfaces or gas phase molecules, as used in the APPI and APLI negative ion mode (cf. Figure 2). In the second step, at atmospheric pressure, the free electrons are readily thermalized via collisions with surrounding buffer gases down to E_{kin} ~ 0.02 eV. Once thermalized, the electrons are captured by molecules that possess positive electron affinities (E_a), forming short lived highly excited molecular radical anions $[M]^{-*}$. This process is significantly enhanced when the electron energy is in resonance with a molecular electronic/vibrational state [59,60]. The fate of the intermediate state $[M]^{-*}$ is described by three feasible reaction pathways: (i) The excess energy is used for bond breaking, which consequently forms a negatively charged fragment $[F_1]^-$ and a neutral fragment $[F_2]$ (electron capture dissociation-ECD), (ii) auto-detachment leads to the reverse of the primary capturing step, and (iii) collisions with a buffer gas take the excess energy of the $[M]^{-*}$ state to form an intact, deactivated radical anion $[M]^-$ [60]. The latter process is called electron capture (EC) or in the case of resonant capturing, resonant electron capture

1 Introduction

(REC). In particular compound classes containing electron density withdrawing functional groups, such as the nitro group, are amenable to (R)EC. However, in many cases direct electron attachment to an analyte is hindered or even not feasible, partly due to low electron affinities, the discrepancy between a resonant state and the kinetic energy distribution of the thermal electrons, or simply due to the occurrence of competing (R)EC processes of other compounds, respectively. To render the negative ion mode more efficient, a secondary ionization method is used. In a first ionization step the abundance of thermal electrons is efficiently converted to charged species of a reactant gas (e.g. O_2), being present in excess. These formed anionic species transfer charge in subsequent reactions onto the analyte. (for more detail see *4.3.3.2 APPI/APLI-Negative Chemical Ionization (NICI)*).

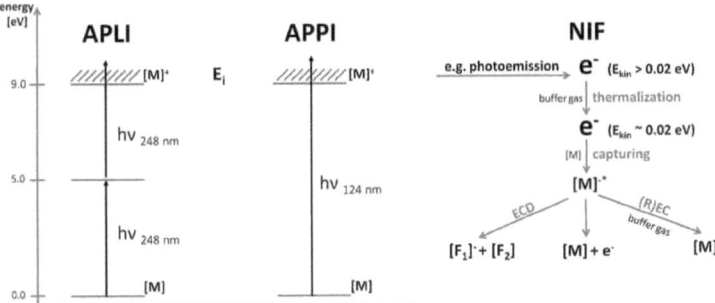

Figure 2: Fundamental processes of APLI, APPI, and NIF. A photon is denoted with hv, and *[M]*, *[M]⁺*, *[M]⁻* and *[M]⁻** denote the neutral, the cationic, the anionic and the excited anionic analyte, respectively. *[F₁]⁻* and *[F₂]* represent anionic and neutral fragments and a non-bound electron is denoted with e⁻.

1.2.1.4 Ion Transformation Processes (ITP)

Any kind of process that transforms the primary generated ion into another structural and/or charged differing species and that consequently changes the resulting ion signal distribution is encompassed by the term "ion transformation process (ITP)".

Accordingly, the rare use of in situ MS methods for degradation product studies in atmospheric chemistry, as noted above (cf. *1.2 Mass Spectrometry*), is due to a general major challenge in mass spectrometry: The fate of charged species between ionization and detection. Effective and unintended transformation processes of the charged species often determine the

1 Introduction

ion distribution at the detection step and hence interfere in the ability of a mass spectrum to reflect the initial neutral composite (cf. Figure 3). These transformations are originating from the strongly changing chemical and physical properties of a neutral molecule with the process of ionization. The concentration of the newly formed charged species generally deviates from any thermodynamic equilibrium. Balancing back to an equilibrium situation is accompanied by molecular structural changes via intra- and/or intermolecular processes. Hereby, the extent of ITPs depends on the chemical properties of the participating compounds, the composition of the matrix, and the number of collisions between ionization and detection such that an API source enclosure should be better treated as a chemical reactor, with a great variety of species present at the same time, to wit: Positive ions, electrons, negative ions, neutral analyte, solvent molecules, neutral radicals, etc. (cf. Figure 3). It is noted that for a complete picture the feasible impact of heterogeneous reactions on the source enclosure walls and within the transfer units to the analyzer should be considered as well.

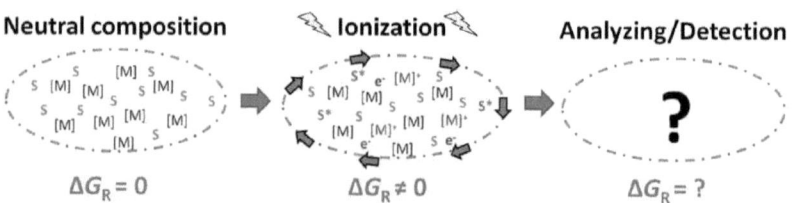

Figure 3: Sketch of the general challenge in mass spectrometry: Does a mass spectrum truly reflect the neutral composition? $[M]$ and $[M]^+$ denote the neutral and the cationic analyte. Matrix compounds and excited or dissociated matrix compounds are represented by S and S*, respectively. Non-bound electrons are denoted with e^-.

Furthermore, potential gradients along the pathway through the MS result in kinetic energy increase of generated ions, which in subsequent collisions enhance decomposition processes. This renders the correct interpretation of MS data even more complicated. However, in several mass spectrometry methods ITPs are desirable and purposively applied. For instance DA-APPI and PTR-MS make use of charge transfer to the analyte by primary generated ions. Furthermore, well known gas phase reactions and purposively evoked dissociation processes,

1 Introduction

such as collision induced dissociation (CID), electron capture dissociation (ECD) and laser induced dissociation (LID) are used to provide valuable structural information. Unintended ITPs, however, basically evoke two crucial aspects for analytical mass spectrometry of *unknown* composites: (i) Ions are detected at significantly differing mass-to-charge ratios than their neutral counterpart accounts for, hence a mass spectrum is affected by "artificial" signals, which are produced by the analytical method itself [61]; (ii) Charge transfer reactions resulting in a thermodynamically controlled ion distribution are leading to loss of MS information in favor of a few dominant species. These two problems are particularly present in API-MS since collision rates of 10^9 s^{-1} are prevailing in the source region. Moreover, partly attributed to polarization effects [62,63], ion-molecule reactions tend to be fast, often exceeding the calculated gas kinetic collision rates when assuming collision cross sections of the corresponding neutral precursors. Considering that the typical API interface may lead to ion dwell times of up to seconds within the elevated pressure region [48], presumably every unbalanced reacting composition has reached thermodynamic equilibrium at the detection step. Besides tentative studies on mechanisms concerning the improvement of DA-APPI or APCI applications [29,64-67], ITPs are poorly discussed in analytical API-MS literature. For example, the impact of the neutral radical chemistry, initiated by electrical discharges (e.g. APCI) or radiation below 200 nm (cf. *1.2.1.2 Atmospheric Pressure Photo Ionization (APPI)*) is basically not discussed. However, it is noted that: (i) At typical APPI and APCI-MS conditions significant *neutral* radical production occurs [53,55,68], and (ii) reactions between ions and neutral radicals are fast [69,70]. As an example the abundant presence of oxygenated aromatic hydrocarbons in API-MS was first mentioned in 1983, when Mahle et al. observed phenoxy type structures upon APCI of benzene [71], however, no evidence for neutral radical chemistry within ion source enclosures was discussed until 2005, when Frey et al. investigated oxidized proteins observed upon APCI [72]. They tentatively explained their observations in analogy to the atmospheric degradation of organic compounds with OH radicals. However, the extent to which neutral radical chemistry is driven in an API source, initiated by VUV light absorption or via electrical discharge processes, has not been recognized yet. Up to now, many reported but unexplained APPI/APCI mass signals are vaguely attributed to impurities or to reactions with closed shell molecules, such as H_2O and O_2. For example, Kauppila et al. [73] and also Robb et al. [67] observed abundant signals of *$[M+15/16/17]^+$* upon toluene DA-APPI. Addition of $CHCl_3$ and H_2O was reported to enhance these signals and revert them to base peaks. These are observations that are readily assigned

1 Introduction

to ITPs with neutral radicals (for more detail see *4.3.2 Neutral Radical Induced ITP (NRITP) [82]*).

Consequently, a more profound understanding of all types of ion transformation processes in API-MS, involving neutrals, cations, anions, electrons, neutral radicals, and also heterogeneous reactions on surfaces is of fundamental interest to avoid significant misinterpretation of MS spectra and to precisely optimize the overall performance of the analytical method.

2 Goals

As mentioned in the introduction there is lack of knowledge in the atmospheric photo oxidation of aromatic hydrocarbons and the need for expanding and improving the analytical instrumentation hitherto used was highlighted. In this work a mass spectrometric system (quadrupole ion trap) is coupled to an atmospheric smog chamber for in situ monitoring of such degradation product studies. A gas phase sampling unit is continuously sampling from the reactor into the ionization region of an atmospheric pressure ionization source. Here, APLI is used to selectively ionize ring-retaining photo oxidation products and possibly secondary organic aerosols with high sensitivity. Furthermore, an integrated APPI setup is supposed to encompass also ring-opened degradation products. The use of these two photo ionization techniques additionally allow for (R)EC, which is particularly sensitive towards nitroaromatic compounds. Additionally, secondary ionization methods, in the positive (also termed as DA-APPI, DA-APLI) as well as in the negative mode further extend the amenable analyte range.

In a first step the usability of the commercially available API source is examined, in particular with respect to ionization and ion transmission efficiencies, based on distribution of ion acceptance (DIA) measurements and fluid dynamical calculations. Hereupon novel approaches in APLI and APPI with regard to source enclosure designs, radiation inlets and radiation sources will be introduced and characterized in detail. Special attention is paid to enhanced technical solutions to better control/reduce ion transformation processes. Also a classification and an extensive discussion on possible ITPs will be given. In consideration of using APPI for the anaylsis of oxidation products, the impact of neutral radical induced ITPs will be discussed in detail, in particular based on reaction pathways of the pyrenyl cation within "VUV-activated" matrices.

Finally an exemplary degradation study will demonstrate the capabilities and limitations of the novel API-MS setup and show effects and parameters that should be considered to obtain valuable structural as well as time dependent information on the occurrence of a degradation product. For more structural information, collision induced dissociation (CID) on selected degradation products will be performed and simultaneously recorded FT-IR data will provide additional qualitative as well as quantitative information on the sample compositions.

3 Experimental

3.1 Mass Spectrometer

All mass spectrometric measurements were performed with an esquire6000 quadrupole ion trap from Bruker Daltonik GmbH, Bremen, Germany (cf. Figure 4, left), originally equipped with a typical API source design (Bruker Apollo™ source) as sketched in section *3.1.2 Common API Source* and discussed in more detail in *4.1 Common API sources*. The working principle of the MS is as follows (cf. Figure 4, right): Ions are generated within the source enclosure and are subsequently forced into the transfer capillary through electrical gradients, created between the spray shield and the metalized entrance of the transfer capillary. The capillary serves as a pressure reduction unit between the atmospheric and the first differentially pumped low pressure region (~4 mbar). An electrical gradient between the metalized end cap of the capillary and the skimmer facilitates the transfer of the ions into the high vacuum pressure region. Ions with a selectable *m/z* range are further guided by the following two octopoles towards an electrostatic lens that focuses the ions into the trap wherein they are accumulated for a certain period of time. Finally, the stored ions are consecutively ejected according to their *m/z* value and guided towards a detector unit consisting of a conversion dynode and an electron multiplier. The generated signal is digitized, normalized to the accumulation time and plotted versus the *m/z* value [74].

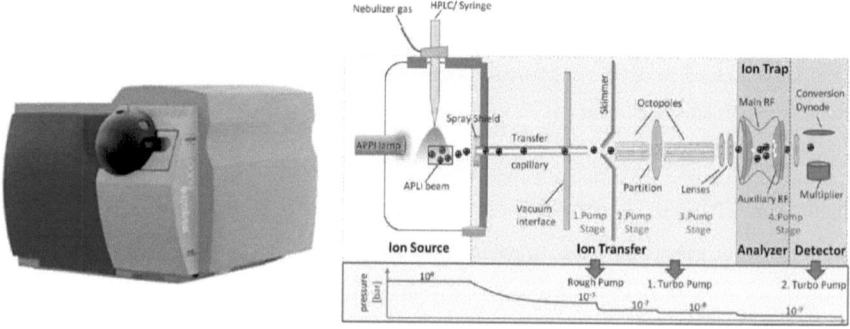

Figure 4: (left) Photograph of the esquire6000. (right) Scheme of the different vacuum stages in the esquire6000, including a pressure diagram.

3 Experimental

The trap is an ion storing device, consisting of two hyperbolic electrodes with their foci facing each other and one hyperbolic ring electrode, separating the two others (cf. Figure 5, left). The entrance cap has one hole, the end cap is perforated. During the accumulation and storing status of the QIT a radio frequency voltage of 781 kHz is applied to the ring electrode [74]. The entering ions are pulled into a pseudo potential well, determined by the so called "trap drive" setting. The periodic motion is composed of two mass dependant individual secular frequencies (ω_r and ω_z), one along the radial (r) and one along the axial (z) direction, respectively. The trapping efficiency is enhanced by a repelling voltage on the end cap and through collisional cooling with a buffer gas (helium), which is leaking into the trap resulting in a pressure of around $5 \cdot 10^{-6}$ mbar. It is noted that the trapping efficiency is significantly affected by the kinetic energy of the ions which they have gained along the upstream guiding potentials. Within an ideal quadrupol field the time dependent trajectory of an ion in the axial direction z and the radial direction r is described by the following equation of motion, originally derived from the Mathieu equation [75,76]

$$m\frac{d^2 z;r}{dt^2} = \frac{-m\Omega^2}{4}\left(a_{z;r} - 2q_{z;r}\cos\Omega t\right)z;r \qquad (eq\ 1)$$

with m as the mass of the ion, t the time, Ω the radial frequency of the RF potential applied to the ring electrode and the dimensionless trapping parameters $a_{z;r}$ and $q_{z;r}$ which are calculated as follows:

$$a_{z;r} = \frac{-8eU}{mr_0^2\Omega^2} \qquad (eq\ 2)$$

$$q_{z;r} = \frac{4eV}{mr_0^2\Omega^2} \qquad (eq\ 3)$$

Here the charge of the ion is denoted with e, its mass with m. U and V represent the amplitudes of the applied DC and RF voltage, respectively, and the size of the trap is represented by r_0. It is noted, that for practical purposes, which will be explained later on, in commercially available ion traps all three electrodes are slightly truncated in order to obtain higher-order multipole contributions to the main quadrupolar field [74]. Consequently, the trapping parameters are modified; however, a mathematical treatment would exceed the scope of this section. The solutions of eq 1 are either (i) periodic but unstable or (ii) periodic and stable, where the latter describe the motion of an ion with a specific m/z value within the dimensions of the trap. Solutions of type (i) determine the boundaries between an unstable

3 Experimental

(ejection in z- or collision with the wall in radial direction) and a stable motion. The characteristic values of the boundaries are described by introducing a new trapping parameter, $\beta_{z;r}$, for each direction. This parameter itself is a complex function of $a_{z;r}$ and $q_{z;r}$ and determines the exact boundary values as integers starting from zero. Commonly $\beta_{z;r}$ values are plotted in a_z and q_z space to obtain the so called Mathieu stability diagram (cf. Figure 5, right). All ions with a_z and q_z values (eq 2 and 3) that are encompassed by boundaries, which are defined through $\beta_{z;r}$ values, render stable, periodic motions within the trap. Since the esquire6000 is operated in the RF-only mode, that is, no DC ($U = 0$) potentials are applied to the electrodes, the trapping parameter $a_{z;r}$ in eq 2 becomes zero and consequently the stability diagram is reduced to a one dimensional line along q_z with the instability boundary at the $\beta_z = 1$-intersection [74].

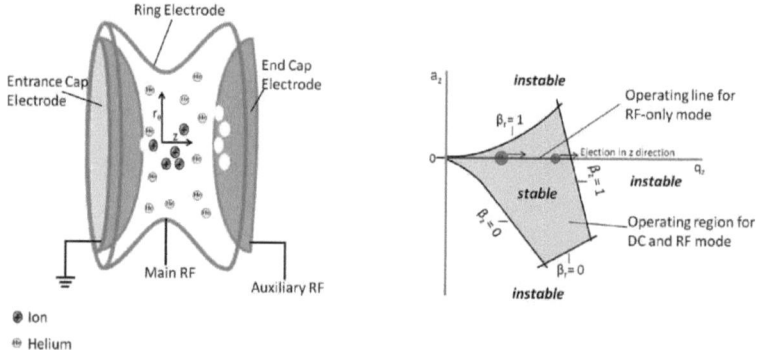

Figure 5: (left) Schematic of the ion trap. (right) Mathieu stability diagram for the ideal quadrupolar field within the trap, generated by the main RF on the ring electrode.

a) Storing mass range capability.

Since in eq 3 the parameter q_z is inversely proportional to the mass-to-charge ratio, lower masses are closer to the unstable boundary than higher masses with the same charge (cf. Figure 5, right). In general this leads to an important consequence for the operation of ion traps: The existence of a lower mass cut-off (LMCO) that is the lower limit of the range of masses that can simultaneously be stored. Theoretically there is no upper trapping limit, since the inverse of the m/z in eq 3 leads to an asymptotical approach of $q_z \to 0$. However, there is experimental evidence that ions with m/z values of 20 - 30 times the lower cut-off mass are

3 Experimental

not efficiently trapped. For the esquire6000 the lowest mass-to-charge ratio that may be stored in terms of software settings is m/z 15, though the actual trapping efficiency remains unknown. Experimental data demonstrate that the trapping efficiency below m/z 50 is significantly decreased. The upper software allowed limit is m/z 3000, or m/z 6000 which is obtained in a special extended operation mode, however, at the cost of resolution [74].

b) Mass discrimination

The discussed confined trapping efficiencies result in significant mass discrimination, which has to be considered when interpreting recorded mass spectra.

c) Mass analysis

The q_z line at $a_z = 0$ is also called the operating line for the mass-selective instability mode, theoretically describing the actual mass analysis of the instrument (cf. Figure 5, right). During the mass analysis the RF amplitude is ramped and, as taken from eq 3, the q_z values of all ions increase with increasing V. Consequently, all m/z values consecutively reach the instability intersection (β_z/q_z) and are one after another ejected in axial direction towards the detector unit. As for the LMCO, the q_z value is inversely proportional to the mass-to-charge ratio; lower masses are closer to the unstable boundary than higher masses with the same charge and are thus ejected first. Additionally to the RF ramp, an auxiliary RF voltage is applied onto the end cap electrode. In combination with the earlier mentioned higher-multipolar fields, a resonant frequency matching the secular frequency ω_z of an ion near the unstable boundary is generated. During the main RF ramp all ions are moved towards the instability boundary along the q_z axis with an increasing secular frequency ω_z. Once this frequency matches the auxiliary and higher-order frequencies, the ions quickly take up energy on their oscillating motion, leading to a significantly enhanced ejection process [74,76].

d) Mass resolution

The resolving power R of the instrument may be defined as

$$R = \frac{m_1}{\Delta m} \qquad \text{(eq 4)}$$

3 Experimental

with m_1 as the mass of a singly charged ion and Δm the FWHM of the Gaussian peak centered at m_1. However, in ion trap mass spectrometry the mass resolution is often specified as simply Δm, since the peak width does not significantly vary with m/z values. The resolving power of a QIT is generally affected by (i) space charge, (ii) the RF ramping speed during the instability scan, and (iii) the kinetic energy distribution of an ion species, which is basically determined by the cooling efficiency of the buffer gas. For the esquire6000 a peak width in the range of $\Delta m = 0.3 - 0.6$ is specified for the mass range m/z 50 - 3000. In case of the above mentioned extended mass range a FWHM peak width of $\Delta m = 5$ is achieved. To account for space charge effects, the instrument offers a so called ion current control (ICC) function that automatically adjusts the accumulation time of the trap to changes in ion concentration [74,76,77].

e) Mass accuracy

The mass accuracy is also affected by space charge contributions and by the kinetic energy distribution of an ion species [77]. The esquire6000 shows accuracy instabilities, even in ICC mode, of $m/z \sim \pm 0.3$. It follows from this discussion that only the nominal is used for data interpretation.

f) Switching positive-negative modus

Since the mass selection in a QIT is solely based on RF potentials, positive and negative ions are treated equally. Hence, this instrument offers a fast alternating modus with switching times of ~ 400 ms between polarities [74].

g) Resonant excitation

As already described for the mass analysis, additionally generated RF fields are used to resonantly excite ions on their trajectories. Therefore, supplementary oscillating potentials, matching either one of the two secular frequencies (ω_r and ω_z) of ion motion, are applied to the end cap electrode. Based hereupon are the fundamental working principles of ion isolation, collision induced dissociation (CID) and sequential mass analysis measurements (Ms^n) [74-76].

3 Experimental

h) Ion isolation

To isolate a narrow range of *m/z* values within the trap, wavebands of frequencies are applied to the cap electrodes, which resonantly excite and thus eject all the undesired ion species [74-76].

i) CID

Following the isolation process, a small frequency band above and below the exact secular frequency of the isolated ion species is applied. This near-resonant excitation promotes the ions into higher potential regions where they gain kinetic energy. Collisions with buffer gas atoms lead to increasing internal energies of the ions and eventually induce bond breaking processes. The higher the applied RF amplitude for the near-resonant excitation the more kinetic energy is gained by the ion. However, the upper limit for this process is given by collision with the trap walls and/or ejection. The trap is capable of holding collisionally generated fragment ions, but only with a lower mass cut off of 27 % (by default) of the parent ion. Thus, for an isolated species with *m/z* 200 only charged fragments down to *m/z* 54 can be trapped and subsequently mass analyzed. At the significant cost of the overall trapping efficiency the default LMCO value can be lowered to 12 % [74]. Relating to the controlled CID experiments are uncontrolled collision induced dissociation processes. Those will inevitably occur at any point of the mass spectrometer when a charged particle gains enough kinetic energy inducing bond breaking after a collision. In ion trap mass spectrometry a critical parameter for unwanted CID is the trap drive, which determines the depth of the potential well with which the ions are forced to move during the accumulation status.

j) Ms^n experiments

One unique feature of ion traps is their capability of performing several subsequent sequences of isolation and CID, also called sequential mass analysis. Hence, structural information is not only provided for the parent ion molecule but also for numerous generations of fragment ions [76].

3 Experimental

k) Duty cycle

The accumulation time may be varied from 0.01 ms up to 50 s and the scan speed can be either 800, 8100 or 26000 $m/z \cdot s^{-1}$. Between accumulation and scan there is a cooling delay and between each scan there is a delay for clearing the trap, both with widths around 5 ms [74]. Thus for a scan speed of 8100 $m/z \cdot s^{-1}$, a mass range of m/z 50 – 500, and an accumulation time of around 50 ms (strongly dependent on the available ion density) a duty cycle for a normal mass scan of around 115 ms results.

l) Chromatogram mode

The esquire6000 was developed for hyphenation with chromatographic stages thus it offers a chromatogram mode in which over a period of time every obtained mass spectrum is stored. Post processing of the raw data than allows for visualization of alterations in time of any m/z trace. This mode is in particular helpful for monitoring degradation product studies.

m) Software

EsquireControl version 6.1 was used as the operating software of the mass spectrometer and post processing of raw data was performed with DataAnalysis 3.4 program package, both from Bruker Daltonik GmbH, Bremen, Germany.

3.1.1 Laser Systems

For APLI studies two OPTex exciplex lasers (Lambda Physik, Göttingen, Germany) radiating at 193 and 248 nm and an ATLEX300 (ATL Lasertechnik, Wermelskirchen, Germany) exciplex laser, also providing 248 nm radiation (cf. Figure 6), were used. Additionally, a wavelength of 266 nm was available (Diode pumped solid state laser (DPSS), FQSS 266-50, CryLas, Berlin, Germany). The working principle of an exciplex laser is based on electrical discharge initiated chemiluminescence within a gas mixture, containing mostly noble gases, halogens, or mixtures of both, along with suitable buffer gases. Frequency adjustable pulsed electrical high voltage discharges (1 – 200 Hz) within Ar/F_2 and Kr/F_2 mixtures initiate a reaction sequence according to the following scheme:

3 Experimental

$$Ar\,(Kr) + e^- \rightarrow Ar^+\,(Kr^+) + 2\,e^- \tag{rxn 1}$$

$$F_2 + e^- \rightarrow F^- + F \tag{rxn 2}$$

$$Ar^+\,(Kr^+) + F^- \rightarrow ArF^*\,(KrF^*) \tag{rxn 3}$$

or

$$Ar\,(Kr) + e^- \rightarrow Ar^*\,(Kr^*) \tag{rxn 4}$$

$$Ar^*\,(Kr^*) + F_2 \rightarrow ArF^*\,(KrF^*) + F \tag{rxn 5}.$$

The subsequently occurring fluorescence step

$$ArF^*\,(KrF^*) \rightarrow Ar\,(Kr) + F + h\nu_{193\,nm\,(248\,nm)} \tag{rxn 6}$$

leads to immediate lasing, since the ground state of the exciplexes is repulsive; upon directly generating electronically excited states, very strong population inversion is obtained. Such systems are called super radiant, since no cavity gain for efficient lasing is required. There is only one high reflector at the end of the laser tube. The pulse duration is typically between 5 and 10 ns for the OPTex lasers and 2 ns for the ATLEX system, with pulse energies in the range of 1 - 8 mJ. The beam profile is of rectangular shape of around 0.5 cm² generating power densities in the order of 10^6 W·cm^{-2} (cf. Table 1).

Table 1: Parameters of the used laser radiation sources.

	ATLEX300; OPTex (exciplex lasers)	FQSS 266-50 (DPSS laser)
Wavelength	248 and 193 nm	266 nm
Pulse energy	1 - 8 mJ	60 µJ
Repetition rate	1 - 200 Hz	200 Hz
Pulse width (FWHM)	5 - 10 ns	1 ns
Beam profile	rectangular	circular
Beam area	0.5 cm²	$2 \cdot 10^{-5}$ cm²
Power density	~ 10^6 W·cm^{-2}	$3 \cdot 10^7$ W·cm^{-2}

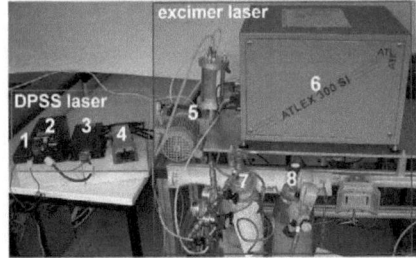

Figure 6: Photograph of the laser systems used. Left: FQSS 266-50 DPSS laser; 1. external 24V-power supply; 2. booster unit; 3. control unit; 4. laser head (can directly be attached to the source enclosure without any further optical devices). Right: ATLEX 300 exciplex laser; 5. vacuum pump with halogen filter; 6. exciplex laser unit (further optical devices required for directing the laser beam into the source enclosure); 7. gas cylinder of the compressed halogen/krypton gas mixture; 8. compressed helium gas cylinder.

The DPSS laser is based on photo excitation of Nd^{3+} ions within a yttrium aluminum garnet matrix (Nd:YAG). Electronic excitation leads into a triplet state with forbidden transition into

3 Experimental

the ground state. The following stimulated emission within a cavity generates laser radiation at 1064 nm. The fundamental is subsequently frequency quadrupled via two consecutive second harmonic generation (SHG) stages. The DPSS laser is operated in a fixed pulse mode (200 Hz) with a pulse width of 1 ns and a pulse energy of 60 µJ. A circular beam profile with an illuminated area of $2 \cdot 10^{-3}$ cm² and a power density of $3 \cdot 10^{7}$ W·cm⁻² is obtained.

Pulse energies were verified with a pyroelectric laser power meter (ORION PE 25-V2, OPHIR Optronics, Jerusalem, Israel).

3.1.2 Common API Source

The API source (Apollo™) originally mounted on the esquire6000 is mainly designed for direct liquid infusion or HPLC-MS applications. The above mentioned multi-purpose ion source (cf. *1.2.1.1 Atmospheric Pressure Laser Ionization (APLI)*) is a slightly modified design allowing for enhanced APLI applications and GC-hyphenation. A principal schematic of the setup along with the prevailing gas flows of these two API source versions is shown in Figure 7 (left). In LC mode the liquid is guided through a needle surrounded by a hot sheath gas (N$_2$, nebulizer gas, 0.5 – 2.0 L·min⁻¹ flow rate), and sprayed orthogonally to the MS inlet into a chamber volume of around 350 cm³. The MPIS enclosure is of rectangular shape, whereas the Apollo™ design is essentially of oval design. Ionization occurs via APPI (Kr-RF, PhotoMate®, Syagen Technology, Inc., Tsutin, CA, USA) or APLI; alternatively, the VUV lamp can be replaced by an APCI needle. The two radiation sources are positioned as shown in Figure 7 (left), with the VUV lamp mounted on the back, slightly upward off axis from the entrance of the transfer capillary. The laser beam is guided through a quartz window, orthogonally to the capillary and the sprayer axes, and positioned 1 – 10 mm in front the spray shield. The generated ions are primarily drawn into the transfer capillary through electrical gradients, which are established by applying voltages onto the spray shield and the metalized end cap of the transfer capillary. The gas stream into the MS is choked[1] and, depending on the dimensions of the capillary (Ø$_{inner}$ = 0.6 mm; l = 18 cm), accounts for Q_{choked} = 1.4 L·min⁻¹ at upstream atmospheric pressure. To enhance evaporation and to prevent solvent drops from entering and possibly blocking the capillary, hot dry gas (0 – 10 L·min⁻¹) is additionally

---

[1] Choking occurs when the gas stream velocity within the capillary reaches sonic speed. At this point the gas throughput cannot be further increased by lowering the pressure on the low pressure side.

3 Experimental

provided which exits through the spray shield, counter propagating the downstream direction of the MS flow. As mentioned above, the MPIS also features GC hyphenation with APLI-MS. In this case, a home-made heated transferline is mounted onto the port of the APPI lamp, and the end of the GC column is placed 2 – 3 cm in front of the spray shield; ionization occurs via APLI as shown in Figure 7 (left) [41,78]. For operation of the source liquid samples were delivered with a syringe pump (Model 100, kd scientific, Holliston, MA, USA) or a Hitachi L-7110 HPLC pump (Hitachi Ltd., Tokyo, Japan) and transferred to the gas phase by pneumatically assisted thermal vaporization, as described. Nitrogen as the main carrier gas was supplied either by a nitrogen generator (UHP LCMS 18, Domnick Hunter, Gateshead, Tyne & Water, UK) or by high purity (99.999 %) nitrogen from compressed gas cylinders.

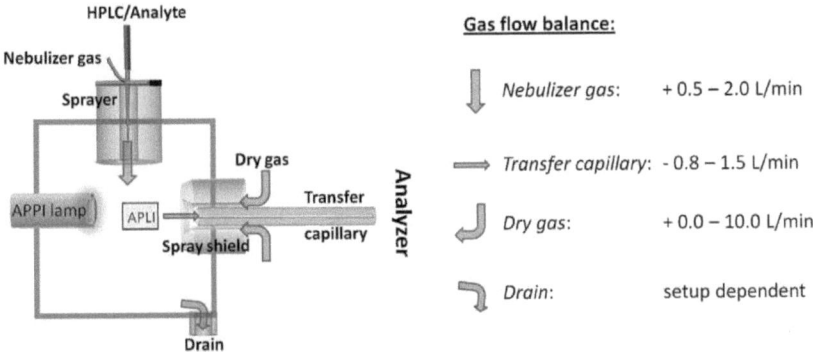

Figure 7: (left) Schematic of the Apollo/ MPIS source [48]. (right) List of the gas flows through the source enclosure. Red and blue arrows indicate the in- and outflows, respectively.

The ozone measurements were carried out with a Thermo Environmental 49 O_3 Analyzer (Thermo Environmental Instrument Corporation, Waltham, MA, USA), connected to the drain of the source enclosure.

3.1.3 Novel Laminar-Flow Ion Source (LFIS)

The laminar-flow ion source is a patented [79,80], home-built API assembly, which was developed within the framework of this thesis. It was first introduced in 2010 by Barnes

3 Experimental

et al. [48]. The source enclosure is made of a metal tube system, primarily designed for ambient pressure gas phase samples (cf. Figure 8). The analyte flow, which is solely determined by the choked flow of the transfer capillary (Q_{choked} = 1.4 L·min^{-1}), enters the sampling tube ($\varnothing_{inner}$ = 9 mm) and is guided at an angle of 10° into the ionization tube of the inlet stage. A subsequent bottleneck reduces the inner diameter to $\varnothing_{inner}$ = 4 mm. The ionization stage is extended to 20 cm and eventually ends in a conically shaped section with an 8° full angle and an orifice diameter of 0.8 mm, which forms a gas tight connection with the following 0.6 mm entrance aperture of the transfer capillary. For APLI applications the laser beam is guided through a quartz window, coaxially into the ion source along the propagation direction of the carrier gas flow. A hole on top of the ionization tube, close to the cone section, allows for light entry. In order to maintain laminar flow, a specially shaped lithium fluoride (LiF) window, in which a flute is cut to match the dimensions of the inner tubing surface, is used. Since the ionization tube is of modular design, i.e., the segments are o-ring sealed (<u>not</u> shown in Figure 8), add-ons and variations in tubing length are possible. The entire assembly can be pumped to roughly 10 mbar in a few seconds by simply shutting off the supply gas. Additionally, thorough cleaning with solvents is easily possible. For APLI the surface of the tubing system is nickel-plated and the reason for this will be discussed in more detail in section *4.2.2.2 LFIS – APLI*. Furthermore, a gauge is connected to the inlet stage to monitor the pressure inside the API source.

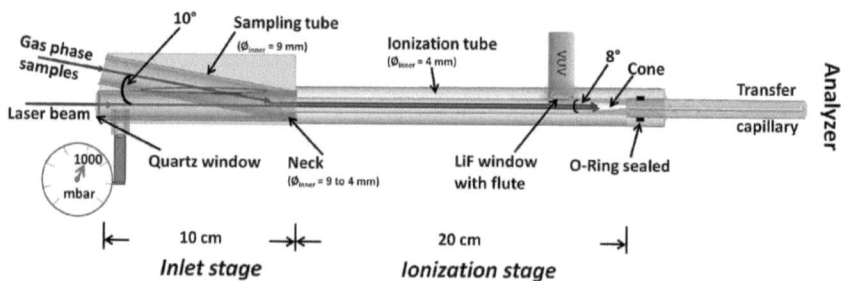

Figure 8: Schematic of the novel laminar-flow ion source.

3 Experimental

3.1.4 Novel APPI Setup

The principle approach of this newly developed APPI setup was first introduced in 2010 [81] and is currently in the patenting process. Within the transfer capillary, functioning as a pressure reduction unit between the atmospheric pressure and the low pressure region of the mass spectrometer (cf. Figure 4, right), a windowless, miniature, pulsed DC-spark discharge VUV lamp is implemented (cf. Figure 9).

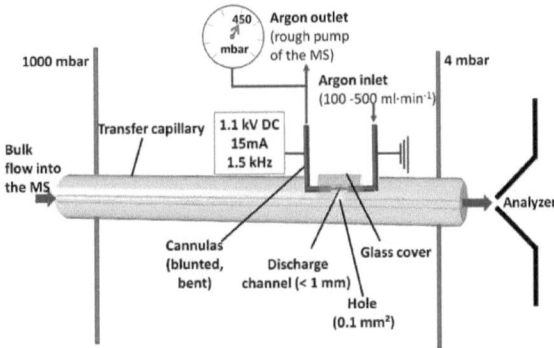

Figure 9: Home-built APPI setup within the transfer capillary separating the atmospheric pressure region and the low pressure region (4 mbar) of the first differential pumping stage.

Several lamp positions were used within this work. The capillary itself is 18 cm long with an outer diameter of $\emptyset_{outer} = 6.5$ mm and an inner diameter of $\emptyset_{inner} = 0.6$ mm. Two conical flutes of 8.0 mm length and 1.0 mm depth are cut into the capillary along the main axis, merging in a hole of 2.3 mm in diameter and 2.0 mm depth (discharge chamber). In the center of the discharge chamber an aperture of around 0.1 mm^2 is cut into the inner tube of the transfer capillary. Two blunted and bent cannulas, serving as electrodes and continuous discharge gas delivery, are inserted into the two flutes with a distance of 0.8 mm to each other, symmetrically placed on top of the aperture. The distance from the discharge region to the bulk gas flow into the MS is about 0.8 mm and the irradiated area 0.1 mm^2. The electrodes, a glass cover on top of the discharge region and the cable for the power supply are all cemented into a small, compact and safe design (Cement NO. 31, Sauereisen, Pittsburgh, PA, USA). A steady argon flow of typically 100 - 500 ml/min is supplied through the grounded electrode,

3 Experimental

which also enhances cooling of the metal. The gas outlet through the opposite high voltage electrode is connected to the rough pump of the mass spectrometer via a needle valve and a bypass to a manometer, allowing for the adjustment of the pressure difference between the discharge chamber and the static pressure inside the transfer capillary. The high voltage (HV) is generated with a HPE CC400 switch-mode power supply of an OPTex exciplex laser, connected to a digital delay generator (Model 9650; Perkin Elmer Inc., Waltham, MA, USA) or a custom designed DD20_10 C-Lader (Hartlauer Präzisions Elektronik GmbH, Grassau, Germany). The frequency is typically set to 1.5 kHz with a trigger signal duration of 0.18 ms.

3.1.4.1 Setup for Characterization of Transfer Capillaries

Some of the fluid dynamical and ion transmission behavior of MS transfer capillaries were investigated with a home-built setup consisting of a vacuum recipient (2 L total volume) which was rough pumped to a background pressure (p_1) between 4 mbar up to atmospheric pressure (cf. Figure 10).

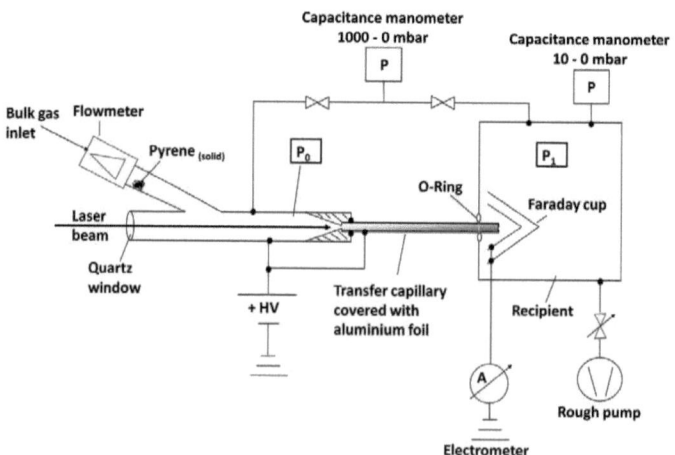

Figure 10: Setup for fluid dynamical and ion transmission characterization of transfer capillaries.

The capillaries were fed into the recipient through a gas tight port. A Faraday cup which is build of two sequentially located conically shaped metal sieves, connected to a Keithley 610C

3 Experimental

electrometer (Cleveland, OH, USA), served as charge detector. The upstream pressure port (p_0) of the capillary was connected to a laminar-flow ion source (cf. Figure 8). Ions were generated with the 248 nm OPTex laser. The gas flow was measured using either a wet meter TG05 (Ritter Apparatebau GmbH and Co. KG, Bochum, Germany) or a FM-360 mass flowmeter (Tylan Corp., Torrance, CA, USA).

3.1.4.2 Characterization of the Discharge Lamp

All spectroscopic measurements in the range 1100 nm - 200 nm were performed with a high resolution fiber optic spectrometer AvaSpec-3648 (Avantes BV, Eerbeek, The Netherlands) connected to a broadly transparent optical fiber (ZFQ-9866, Ocean Optics, Dunedin, FL, USA). Scans between 200 nm and 105 nm were accomplished with an ARC VM-502 VUV spectrometer (Acton Reasearch Corporation, Acton, MA, USA) equipped with a MgF_2 coated parabolic grating (1200 G·mm^{-1}). A 2 mm thick, optically polished LiF window (2045120f, Korth, Altenholz, Germany) was mounted in front of the radiation entrance slit and a home-built cascaded microchannel plate detector unit served as VUV radiation detector at the exit slit. The electrical signal was monitored with a digital multimeter (VA18B, Shanghai Yihua V&A Instruments CO.,LTD, Shanghai, China) and transferred at a rate of 2 Hz to a personal computer as time/signal data points. The actual spectrum was obtained in a post processed time/wavelength conversion. The scan speed was set to 0.1 nm·s^{-1}. With 10 µm wide entrance and exit slits a resolution of 1.7 nm (FWHM) was obtained. The VUV spectrometer chamber pressure was held at 10^{-4} mbar with a turbo- and membrane pump assembly. Due to the incompatibility of sampling VUV spectra within the transfer capillary a special design of the miniature spark discharge lamp was used for the VUV spectroscopy experiments with basically identical dimensions and discharge characteristics as the capillary mounted lamp.

The temporal evolution of a single spark in the UV/VIS was investigated with a photodiode (SD 200-12-22-041, Advanced Photonics, Inc., Camarillo, CA, USA), which was operated in reversed-biased mode and connected to an oscilloscope (TDS 1012, Tektronix, Inc., Wilsonville, OR, USA). Visualization and data processing was done with a self-programmed viewer based on LabVIEW 7.1 (National Instruments, Austin, TX, USA). The oscilloscope was also used for time dependent voltage and current investigations of a

3 Experimental

single spark event. For this purpose a high voltage probe was connected and the voltage drop across a known resistance (0.6 Ω) between the anode and ground was measured, respectively.

3.1.3 Setup for Neutral Radical Induced ITP Studies [82]

The home-built ion source (cf. Figure 11) was designed for operation with gaseous compounds only and it was first introduced by Kersten et al. [82]. It consists of a 25 cm long quartz tube, with an inner diameter of 8 mm, directly attached and sealed to the transfer capillary of the mass spectrometer. A pyrex tube of $\varnothing_{inner} = 4$ mm, entirely opaque for VUV and UV light is coaxially placed in the center of the quartz tube and ends 6 cm upstream of the MS transfer capillary. The 193 nm (VUV) laser beam is directed 5 mm upstream of the glass tube and the 248 nm (UV) laser beam 5 mm downstream of the glass tube exit. In this way the analyte, delivered through the inner glass tube, is not exposed to the VUV radiation, but ionized with UV radiation. However, the VUV laser is able to produce significant amounts of *neutral* atoms and radicals upon photolysis of precursors such as O_2, H_2O, CH_nCl_{4-n} (n=0 – 2), which are delivered in a controlled fashion within the bulk gas flow surrounding the inner glass tube. Hence, this setup spatially separates the 2-photon UV laser ionization region of the analyte from the production zone of neutral radicals by 1-photon VUV laser photolysis. The total gas flow is held constant at 1.4 L·min^{-1}. This closely matches the experimentally determined choked flow through the transfer capillary. The mean gas flow velocity through the quartz reactor tube is calculated to be 46 cm·s^{-1}, resulting in a Reynolds number of $Re = 210$. Hence, the transport through this setup is essentially laminar, allowing for rather accurate calculations of the residence times and thus reaction times. Additionally, this setup can be pumped down to roughly 1 mbar in a few seconds simply by shutting off the supply gases. The gas flows are regulated either by the inbuilt flow controller of the MS and/or by external mass flow controllers (1179A Mass-Flo-Controller, 2000 sccm, mks instruments, Andover, MA, USA) connected to a 647 C Multi-Gas-Controller unit (mks instruments, Andover, MA, USA). Liquid compounds are transferred to the gas phase using stainless steel cryo-traps, in which the saturation vapor pressure of a compound at constant temperature is balanced with nitrogen to a total pressure of 3000 mbar. The flow controllers allow for an accurate delivery of the compounds to the carrier gas. The final concentration is calculated from the known vapor pressure at the prevailing room temperature, the mixing ratio after pressurizing with nitrogen and the mixing ratio within the total flow. The delivery

3 Experimental

of gaseous pyrene is accomplished by equilibrating the saturation vapor of deposited solid pyrene on the walls in the tube system with the main gas flow. This results in a constant pyrene mixing ratio of max. 6 ppbV (parts per billion by volume), calculated using a pyrene saturation vapor pressure of 610^{-6} mbar at 25°C [83] and a total pressure of 970 mbar in the ion source.

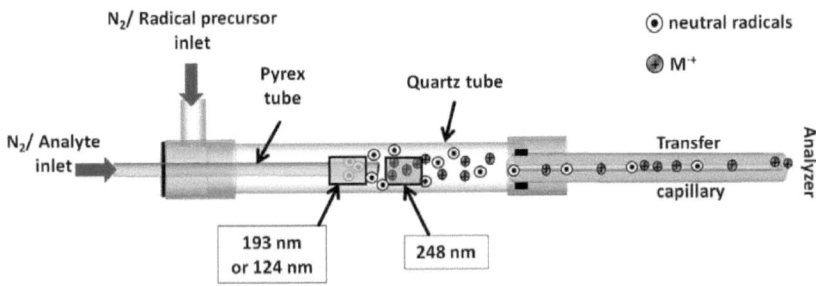

Figure 11: Home-built ionization setup for kinetic and mechanistic studies [82].

3.3 Photoreactor

The gas phase degradation experiments were performed in a 1080 L quartz glass reactor with 6.2 m length and an inner diameter of 0.47 m [15,16]. A turbo-molecular pump backed by a double-stage rotary rough pump allows for evacuation to 10^{-4} mbar. The chamber is surrounded by 32 superactinic fluorescent lamps (320 nm < λ > 480 nm) and 32 low pressure Hg-lamps (λ_{max} = 254 nm) which are individually controlled. Homogeneous mixing of the reactants inside the reactor is ensured by three magnetically coupled Teflon fans. Both ends of the chamber are closed by aluminum flanges and allow for versatile connections (cf. Figure 12: 1).

3.3.1 Procedure of Atmospheric Degradation Studies

For typical OH radical initiated degradation experiments around 1 ppmV (parts per million by volume) of the compound that was to be investigated, 2 - 5 ppmV nitricoxide (NO)

3   Experimental

and 2-5 ppmV methylnitrite (CH_3ONO) were injected into the photoreactor and backfilled with synthetic air to atmospheric pressure. Switching on the superactinic fluorescent lamps induced the production of OH radicals via photo dissociation of CH_3ONO followed by a subsequent reaction cascade with O_2 and NO:

$$CH_3ONO + h\upsilon\ (\lambda > 360\ nm) \rightarrow CH_3O\bullet + NO\bullet \qquad (rxn\ 7)$$

$$CH_3O\bullet + O_2 \rightarrow HCHO + HO_2\bullet \qquad (rxn\ 8)$$

$$HO_2\bullet + NO\bullet \rightarrow OH\bullet + NO_2\bullet \qquad (rxn\ 9)$$

The experiment was typically stopped after 30 to 40 minutes with 50 to 60 % degradation of the initial compound. Considering starting concentrations of around 1 ppmV, oxygenated products in the ppbV range were thus expected.

3.3.2 FT-IR-setup

The reactor is equipped with a Nexus Fourier transform infrared (FT-IR) spectrometer (Thermo Nicolet Corp., Madison, WI, USA), using a liquid nitrogen cooled mercury-cadmium-telluride detector. A White-type multiple-reflection system with a base length of 5.91 m gives a total optical path length of 484.7 m at 82 traverses. The IR spectra encompassed the spectral range of 4000 – 700 cm^{-1} with a resolution of 1 cm^{-1} (FWHM).

3.3.3 MS Sampling Unit

For continuous sampling into the MS, the ion trap is connected to the photoreactor as follows: A glass tube (Figure 12: 2) of 20 cm length with an inner diameter of 8 mm extends into the reactor and ends in a gate valve (Figure 12: 3) on the outside of the reactor, mounted on the aluminum flange. The transfer to the laminar-flow ion source (Figure 12: 6) is accomplished via a second glass tube (Figure 12: 5) of the same dimensions, surrounded by an aluminum housing, allowing for additional sheath gas flow (Figure 12: 4). The continuous sample flow from the reactor is determined by the already described choked flow of the MS transfer capillary (Q_{choked} = 1.4 L·min^{-1}). Considering the dimensions of the entire tubing system a sample needs roughly 0.9 s from the reactor to the entrance of the MS. The pressure in the reactor is balanced to ~1000 mbar by providing a continuous flow of synthetic air matching the amount sampled by the MS. The resulting dilution in the chamber amounts to 8 % after 60 min.

3.3.4 MS Ionization Unit

The above described laminar-flow ion source (cf. *3.1.3 Novel Laminar-Flow Ion Source*) is used for atmospheric pressure ionization. The integrated APLI setup provides an optical beam delivery stage that directs the laser beam of the DPSS laser (cf. *3.1.1 Laser Systems*) coaxially into the ion source along the propagation direction of the bulk gas flow (Figure 12: 11). The novel APPI setup within the transfer capillary (*cf. 3.1.4 Novel APPI Setup*) additionally allows for single photon ionization (Figure 12: 9). The entire laminar-flow ion source (Figure 12: 6) in connection with the transfer unit from the reactor (Figure 12: 5) is pumped to roughly 10 mbar in a few seconds by simply shutting off the supply gas with the gate valve (Figure 12: 3), facilitating frequent flushing of the system.

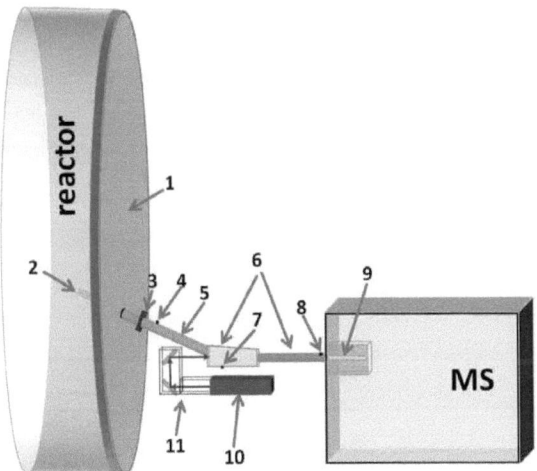

Figure 12: Schematic of the experimental setup used for degradation studies. 1) aluminum flange of the reactor; 2) sampling glass tube; 3) gate valve; 4) sheath gas inlet; 5) transfer unit (glass tube, surrounded by an aluminum housing for sheath gas supply); 6) laminar-flow ion source; 7) port for pressure measurement; 8) VUV radiation inlet stage; 9) transfer capillary unit with the home-built APPI setup; 10) DPSS laser; 11) laser optics (mirrors).

3 Experimental

3.4 Chemicals

Acetonitrile (ACN), methanol, acetone and tetrachloromethane were obtained from Fisher Scientific, Waltham, MA, USA. Benzene, anthracene and pyrene were purchased from Merck KGaA, Darmstadt, Germany. The xylene isomers (p-,m- and o-) were from Fluka, Buchs, Switzerland. Nitrogen oxide was from Air Liquide Deutschland GmbH, Düsseldorf, Germany, with a stated purity of 99.5 %. Oxygen, argon; nitrogen, helium and synthetic air were from Gase.de, Sulzbach, Germany, with a stated purity of 99.999 % vol. A compressed gas cylinder with a mixture of 17 oxygenated VOCs in synthetic air was generated in house. The mixing ratios ranging between 10 and 220 ppbV, were confirmed through a calibrated GC-MS analysis. Methylnitrite (MeONO) was synthesized as described in [84] and stored at -78 °C. All solvents were of analytical or chromatographic purity, all other chemicals of highest purity available, and were used without further purification.

3.5 Computational Investigations

Investigations on neutral radical induced ion transformation processes were supported by kinetic simulations with the Chemked-I/II version 3.3 software package [85].

Gaussian 03W [86] combined with the graphical user interface GaussView 4.1 [87] is used for potential energy calculations along the proposed reaction pathways in the section of neutral radical induced ion transformation processes. The reported Gibbs free energies and enthalpy corrected total energies are obtained from geometry optimizations and subsequent frequency calculations using the Density Functional Theory (DFT) with the Becke-3-Parameter-Lee-Yang-Parr functional (B3LYP) and the 6-31++G(d,p) doubly-diffuse and doubly-polarized split-valence basis set. The combination of this functional and basis set type provides the optimal cost-to-benefit ratio with respect to CPU time [88] and has satisfactorily been applied to a large number of comparable reaction systems [89-92].

Ion trajectory simulations were performed with the SIMION™ Version 8.0 software package [93] and for fluid dynamical simulations Comsol Multiphysics Version 4.0a (Comsol AB, Stockholm, Sweden) was used.

4 Results and Discussion

4.1 Common API sources

Commercially available API sources that are offered by the leading mass spectrometry manufacturers such as Bruker Daltonics with the Apollo™ series, Thermo Scientific with the Ion Max™ source or Waters with the Z-Spray™ or Xevo™ line are all similar to the basic design shown in Figure 7 [94,95]. The following working principle (cf. *3.1.2 Common API Source*) holds true for all these setups: (i) Pneumatically assisted thermal vaporization of a liquid flow into a chamber of considerable volume, (ii) ionization within this chamber and (iii) presence of electrical potential gradients draw the ions into the MS. The vaporization step results in significant gas input, since there is an additional sheath gas assisting the formation of the spray and a gas stream counter propagating the direction of the MS flow to prevent droplets from entering the mass spectrometer (cf. Figure 7, right). The flow through the MS-inlet, when operated with a transfer capillary as a pressure reduction stage, is considered to be nearly constant, as long as there is no significant pressure increase or temperature change on the atmospheric pressure side (cf. *4.2.3.1 Characterization of Transfer Capillaries*). The least defined volume flow is at the drain. Its value is basically laboratory dependent, some instruments are left open towards the atmospheric pressure and some are connected to an exhaust vent installation, which is held at slightly decreased pressure (cf. Figure 7, right). Consequently, complex fluid dynamical and ion transmission behavior for this type of source design is expected.

4.1.1 Distribution of Ion Acceptance (DIA) Studies

In 2007 Lorenz et al. [43] started in-depth investigations on the characteristic transport processes and basic ionization characteristics that are prevailing in this type of API assembly. Their approach to obtain further insight into these processes is based on spatially and temporally resolved APLI experiments within the Apollo™ source, the Z-Spray™ source, and the MPIS. Briefly, an area of $14 \cdot 14$ mm², spanned by the axes of the MS sampling port and the vaporization stage is scanned by a focused laser beam. The recorded mass spectrometric signal is then plotted against the laser position (cf. Figure 13) in a two-dimensional, color

4 Results and Discussion

coded diagram. The thus obtained "distribution of ion acceptance (DIA)" plots show the two-dimensional relative distribution of the sensitivity within the source enclosure. DIA measurements reveal the region of the source enclosure in which ions can be produced and are detectable by the MS. It is noted that such a plot depicts the product of two basically orthogonal parameters: (i) The spatial overlap of the neutral analyte density distribution with the laser beam irradiated volume (affecting the *ionization* efficiency), and (ii) the overlap of the resulting ion population with the spatial distribution of ion acceptance of the MS (affecting the *in-source detection* efficiency). Both effects depend on the laser beam position and on the gas flow dynamics; the in-source detection efficiency is furthermore influenced by electrical potential gradients. Figure 13 depicts an exemplary DIA for the MPIS with common ion source parameter settings as used for routine LC-APLI.

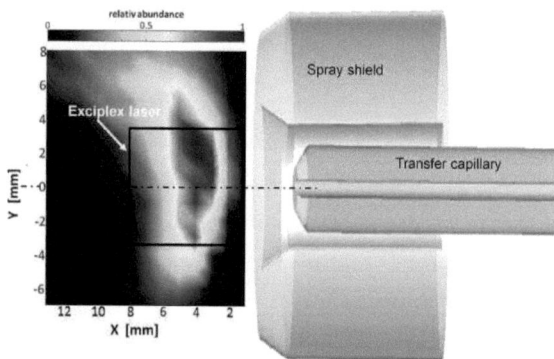

Figure 13: DIA plot obtained with common ion source parameter settings for routine LC-APLI with the MPIS [44]. For comparison the cross sectional dimension of an exciplex laser beam is sketched.

Such a sensitivity distribution requires strong electrical fields facing toward the transfer capillary and voltage settings of typically -1000 V at the capillary entrance are applied. The projection of the two-dimensionally resolved experiments shown in Figure 13 is of ellipsoidal shape with a vertical extension of 14 mm, a horizontal extension of 3 mm and an area of 35 mm². The DIA is located 3 mm in front of the spray shield, located nearly symmetrically around the horizontal axis of the capillary entrance [44]. DIA measurements have proven to

4 Results and Discussion

be a powerful tool for investigations on temporally and spatially resolved ionization and ion transmission performance of API sources [43,46,47,96].

One of the most important findings based on DIA data is that the region of maximum overlap of the ionization and in-source detection efficiency is rather confined. It follows that the spatial extension in which primary generated ions are detected by the MS is small compared to the enclosure volume. This entails two crucial aspects: (i) The ionizing unit (laser beam) can only be positioned in a narrow volume of the total enclosure to obtain optimal performance and (ii) the position of this volume is significantly affected by the prevailing gas flows and applied electrical potentials. It is noted that current DIA data cannot be directly related to the APPI setup, since an additional repelling voltage is applied to the housing of the VUV lamp, generating a potential gradient between the spray shield and the lamp. Such DIAs have not been investigated yet, but rough estimates lead to the conclusion that the shape, dimension and dependency on source parameters of such a sensitivity distribution should not significantly differ from the hitherto obtained results.

4.1.2 Fluid Dynamical Behavior

In 2010 a computational fluid dynamic (CFD) approach, experimentally validated by particle image velocimetry (PIV) measurements, was introduced to gain more insight into the prevailing mechanisms of the common API sources [48]. The computed models fully proved the hypotheses hitherto solely based on DIA measurements and furthermore significantly extended the current knowledge base. In Figure 14 three important aspects are presented: (i) The velocity distribution (left), (ii) the time integrated trajectories (center) and (iii) the neutral analyte distribution (right). As can be seen in the velocity distribution the main gas drift is determined by the gas stream from the sprayer with nearly 8 $m·s^{-1}$ perpendicular to the MS inlet. The trajectories shown slow down to 2 $m·s^{-1}$ before they strike the enclosure bottom and split with a subsequent upward directed circular motion and further deceleration to less than 1 $m·s^{-1}$. The time integrated trajectories show analyte dwell times in the order of seconds. Transferring this to the neutral analyte density variation model a nearly isotropic distribution within the entire enclosure results. Consequently, these computed models reveal rather chaotic, turbulent, most probably strongly instationary flows causing long residence times within such a source design. This inevitably enhances uncontrollable hetero- and homogenous

4 Results and Discussion

neutral- and ion transformation processes, ion losses, memory effects, and chromatographic peak broadening. Time resolved mechanistic studies of unknown compositions, such as the intended atmospheric degradation experiments, are rendered nearly impossible, since the impact of the analytical method itself on the primary generated ion distribution is an unspecified parameter.

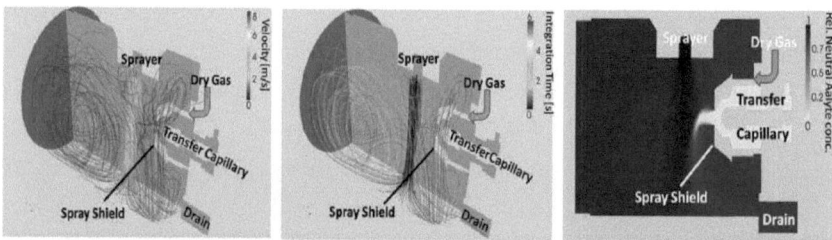

Figure 14: Computational fluid dynamical calculation results (Ansys CFX-11) of (left) the velocity distribution, (center) time integrated trajectories and (right) the neutral analyte distribution within the geometries of a MPIS based on typical source settings [48].

In typical LC-APPI measurements the penetration depth of VUV light is less than 5 mm (cf. *1.2.1.2 Atmospheric Pressure Photo Ionization (APPI)*), thus the APPI lamp is surrounded by a roughly calculated illuminated volume of 0.07 cm^3 (considering a nearly spherical shape). With respect to the confined DIAs it becomes apparent why poor direct photo ionization efficiencies have been reported [51]. First of all, the overlap of the neutral analyte density and the irradiated volume is small, since prior to the ionization step an extensive neutral analyte distribution occurs. Secondly, the overlap of the resulting ion population with the spatial distribution of ion acceptance of the MS is very sensitive to ion source settings and changes in fluid dynamics. Consequently, the only meaningful way to perform efficient APPI in the present source design is via a dopant that directs the light energy into ionization processes. In this case, long residence times and an isotropic distribution of dopant ions is desirable, since subsequent chemical processes such as charge transfer to or protonation of the analyte are enhanced.

The UV laser light in APLI applications, however, is not restricted to absorbing matrix components along the propagation direction. Here the ionization volume is merely restricted to the dimensions of the conventional assembly. Together with the rather confined overlap of

4 Results and Discussion

the resulting ion population and the spatial distribution of ion acceptance of the MS the ineffective use of the ionizing laser radiation becomes apparent [44].

Furthermore, observations of significant signal variability induced by external "parameters" (window open, door shut) have been made with the conventional source design. Qualitatively, these effects are attributable to changes in the fluid dynamics, which might significantly shift the DIA within an experiment hence causing alternating signal intensities in the obtained mass spectrum.

In summary, thorough studies of the fluid dynamical and ion transmission behavior of the commercially available atmospheric pressure ionization source [48,97,98] revealed chaotic, turbulent flows and long residence times within the source enclosure. Uncontrollable hetero- and homogenous neutral- and ion transformation processes, ion losses, memory effects, and chromatographic peak broadening are enhanced and constitute an unspecified operational parameter, which may severely impact on the source performance. Thus the reliability of MS data recorded with these types of ion sources is in fact questionable, rendering mechanistic studies with unknown compounds nearly impossible. The conventional assembly additionally restricts the ionization volume and leads to an insufficient use of the ionizing laser radiation [44]. Hence, the application of this type of source design was deemed inappropriate for the intended gas phase sampling from atmospheric degradation product studies.

4.1.3 H_2O and O_2 Background Concentrations [82]

Usually a factory delivered API mass spectrometer has a nitrogen generator to provide a continuous carrier gas supply without the need of frequently replacing compressed gas cylinders. In this section the permanent background concentration of water and O_2, inevitably supplied by such a generator in varying amounts and thus being present in the Apollo™ source was investigated. The latter was experimentally determined by measuring the ozone mixing ratio as a function of added known amounts of pure oxygen to the main gas flow in the presence of 193 nm laser radiation. The oxygen is partly photodissociated upon irradiation, which results in increasing concentrations of ozone primarily due the reaction sequence:

4 Results and Discussion

$$O_2 + h\nu \rightarrow 2\ O(^3P) \qquad \text{(rxn 10)}$$

$$O(^3P) + O_2 + M \rightarrow O_3 + M \qquad \text{(rxn 11)}$$

(at low water mixing ratios the secondary photolysis of O_3 results essentially in a null cycle; cf. reactions No 1/5/6/7 in Table 4). Plotting the measured ozone mixing ratio as a function of the added oxygen yields a straight line, which is extrapolated to the zero point of the ordinate yielding an O_2 mixing ratio of 2.4%, as shown in Figure 15. This procedure merely represents the well known analytical method of standard additions.

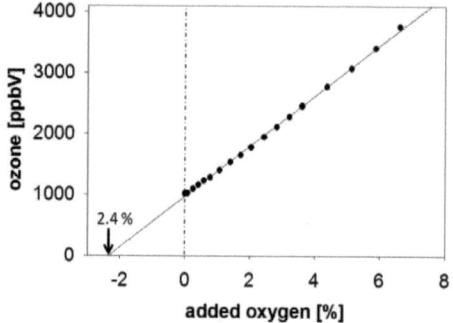

Figure 15: Standard addition method to determine the oxygen mixing ratio in the present API source. The ordinate intercept of (-)2.4 % represents the source background O_2 mixing ratio [82].

Routine operation of API sources, e.g., frequent solvent and sample change, frequent opening of the source enclosure for maintenance purposes, leads to highly variable water mixing ratios in the source enclosure. With respect to solvent purity, the range of dissolved H_2O varies from 100 ppm in non-polar solvents, such as heptane or hexane, up to several percent in more polar solvents, e.g., CH_3OH or CH_3CN. Here, only the water background concentration within the source, when all bulk gases are flowing was determined. A flow controller at the exhaust line of the ion source was used to maintain a constant gas flow. The exhaust gases were passed through a stripping coil, which was held at −78 °C, leading to a quantitative condensation of gas phase water. By re-weighing the mass of the coil the water background mixing ratio in the gas flow was determined to be roughly 100 ppmV. The water

ated after flushing the source for 20 hours with heated nitrogen (350 °C), obtained from the nitrogen generator of the MS. The manufacturer states an initial N_2 purity with less than 0.5 % oxygen present. At the time of the experiments, the built-in cartridge was about 11 months old. There is no information on the effectiveness of this cartridge with respect to reduction of other compounds such as H_2O.

4.2 Development of a Novel API Approach

The above investigations on the fluid dynamic behavior and ionization capabilities of the conventional API assembly pointed out the need for a more controllable system, tailored to gas phase sampling (in a first approach), efficient VUV ionization and the efficient use of the laser beam in APLI applications.

4.2.1 LFIS - Preliminary Experiments

a) Rough determination of the flow characteristics

Preliminary tests have been performed with tubing systems as ionization source enclosures that were directly mounted onto the transfer capillary. The flow within these systems was determined by the choked flow of the transfer capillary, no potential gradients for ion deflection purposes were applied and the entire bulk gas flow was delivered into the mass spectrometer. The dimensions of the tubes ranged from $l = 50$ cm down to 20 cm length, with inner diameters of $\varnothing_{inner} = 4$ mm up to 9 mm. The Reynolds numbers were calculated as follows:

$$Re = \frac{\rho \cdot v_x \cdot \varnothing_{inner}}{\chi} \qquad (eq\ 5)$$

with $\rho = 1.17$ kg·m^{-3} as the density of nitrogen at 293 K, v_x [m·s^{-1}] as the averaged axial drift velocity (0.4 – 1.9 m·s^{-1}), which is determined by the tube dimension and the choked capillary flow ($Q_{choked} = 1.4$ L·min^{-1}), $\varnothing_{inner}$ [m] as the tube diameter and $\chi = 1.76 \cdot 10^{-5}$ kg·m^{-1}·s^{-1} as the dynamic viscosity of N_2 at 293 K. The obtained Reynolds numbers range from $Re = 249$ up to $Re = 559$, well below the critical number of $Re_{crit} = 2300$ [99], consequently essentially laminar flows prevailed in all these tubing systems. In Figure 16 [48] a computational fluid

4 Results and Discussion

dynamical model of such a tube is shown. The boundary conditions were $Q_{choked} = 1.4$ L·min^{-1}, $Ø_{inner} = 9$ mm and a conically shaped end with 0.8 mm orifice. The solid lines represent stream lines, the size and direction of arrows indicate the velocity components and quantitative information of the prevailing velocity is given by the colored scale. The simulation clearly reveals an essentially laminar flow without ablation from the main stream or formation of turbulences close to the cone region. Ion losses are expected to occur mainly by diffusion to the wall.

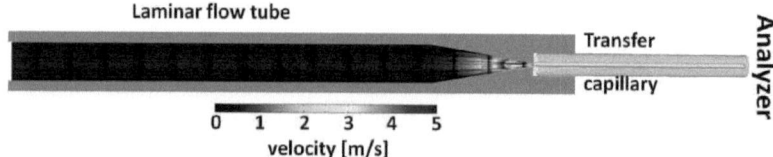

Figure 16: Fluid dynamical simulation of a tube with a volume flow of 1.4 L·min^{-1}, inner tube diameter of 9 mm and an 0.8 mm orifice at the segue to the transfer capillary [48].

b) Ion transmission efficiencies

In one of the first approaches, as shown in Figure 17 (left) [48], a quartz tube of 40 cm length and an inner diameter of $Ø_{inner} = 4$ mm was connected to the capillary entrance.

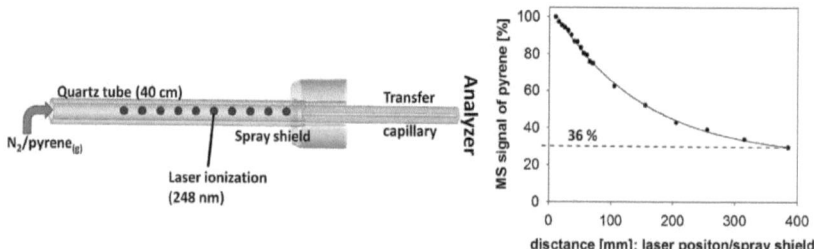

Figure 17: Preliminary test on ion transport efficiencies in a laminar flow [48]. (left) Scan of ionization positions with a 248 nm exciplex laser (beam collimated to 1.5 mm cross section) along a quartz tube of 40 cm length. (right) The obtained relative mass spectrometric signal is plotted against the laser spot distance to the spray shield [48].

4 Results and Discussion

The end of the tube was sintered to approximately 0.8 mm in diameter ensuring a smooth transfer into the capillary. The delivery of analyte was accomplished by equilibrating the saturation vapor of deposited solid pyrene on the walls in the tube system with the main gas flow of nitrogen. An aperture plate with holes of 1.5 mm in diameter was positioned outside along the tube main axis. The holes were subsequently irradiated with a 248 nm laser radiation (100 Hz and 3.5 mJ per pulse accounting for 23 µJ per pulse within the tube). It is noted that the mass spectrometric signal was very sensitive to small alterations of the laser spot position perpendicular to the tube axis. In Figure 17 (right) the obtained signal for pyrene is plotted against the laser beam position. It shows that 36 % of the signal intensity remains even after 40 cm of transport. Simultaneous illumination of several spots with an expanded laser beam revealed that the recorded signal corresponds well to the sum of signal intensity observed individually at each position. For a rough quantification a function is fitted to the plot in Figure 17 (right) and the following equation is obtained:

$$I = 23.8 + 83.0 \cdot e^{-0.0069d} \qquad (eq\ 6)$$

with I as the relative mass spectrometric signal of pyrene in [%] and d in [mm] as the distance of the illuminated spot from the spray shield. For a calculated estimate of the signal strength generated with a coaxial laser beam that is positioned along the propagation direction of the bulk gas flow, an integrated form of eq 6 is required:

$$[\Delta I]_{d_1}^{d_2} = \left[23.8 \cdot d - 83.0 \cdot \frac{e^{-0.0069d}}{0.0069} \right]_{d_1}^{d_2} \qquad (eq\ 7).$$

Considering a tube of 20 cm length which is coaxially illuminated with a laser beam of the same operating conditions as used for the experiment in Figure 17, with $d_1 = 0$ mm and $d_2 = 200$ mm, an integrated relative intensity of $1.3 \cdot 10^4$ % would be obtained. Compared to the maximum of 100 % (cf. Figure 17 right) with the perpendicular single spot illumination an at least 130-fold increase is expected for a coaxial configuration. In addition, the ionization efficiency is enhanced due to multiple illumination of the neutral analyte. The volume inside the tube can be envisioned as an infinite number of connected segments (neglecting the Hagen-Poiseuille flow profile), each propagating with the average drift velocity $v_x = 1.9$ m·s^{-1} (with $Ø_{inner} = 4$ mm, $Q_{choked} = 1.4$ L·min^{-1}) towards the entrance of the capillary. With a tube length of $l = 20$ cm the dwell time of such a segment accounts for 0.1 s. Considering the laser repetition rate of 100 Hz one segment is illuminated ten times before entering the transfer

4 Results and Discussion

capillary. In the conventional perpendicular APLI operation mode (cf. *3.1.2 Common API Source*), however, a neutral analyte molecule is irradiated at most only once [44].

c) Different behavior of quartz and metal tubes

Similar investigations on stainless steel tube systems revealed comparable results in terms of ion transport efficiencies. Significantly differing behavior, however, was observed with respect to response times, here referred to as the delay between the start of the delivery of ionizing radiation and recording a stable mass spectrometric signal. Perpendicular ionization within a metal tube system readily led to the expected response times determined by the average axial drift velocity v_x. In contrast, the characteristics in quartz tubes were rendered much more difficult, since significant charging effects were observed. In this way a quartz tube that had not been used for a longer period of time exhibited response times of up to 10 minutes with a perpendicular ionization positions of $d = 40$ cm. Once being "conditioned" the response time of a tube reproducibly followed the expected average axial drift velocity. Furthermore, applying partially located electrostatic fields onto a quartz tube, e.g. touching with a finger, caused immediate collapse of the MS signal, however, after a certain period of time the signal recovered. No such effects have been observed with metal tube systems. Apparently, it is possible to induce partially located strong potential gradients within a quartz tube that are strong enough to significantly affect the ion transport. On the contrary, the equipotential surface of a metal tube ensures field-free conditions.

d) Impact of the laser frequency in coaxial configuration

Here the setup for capillary investigations as shown in Figure 10 was used with an ionization tube of $\varnothing_{inner} = 9$ mm and 20 cm length. The gas flow was 1.5 L·min^{-1} resulting in an average drift velocity of $v_x = 0.4$ m·s^{-1} and a dwell time of 0.5 s. The 248 nm laser beam was reduced to a circular profile of 6 mm in diameter with pulse energies of 0.7 mJ. Pyrene was delivered as shown in Figure 10, accounting for an estimated neutral analyte concentration of around 6 ppbV (cf. *3.1.3 Setup for Neutral Radical Induced ITP Studies [82]*). Figure 18 depicts the recorded ion current, plotted as function of the laser frequency and the number of segment illumination. It clearly demonstrates significant enhancement of signal intensity due to multiple segment illumination.

4 Results and Discussion

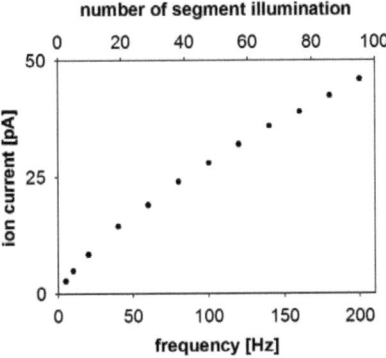

Figure 18: Recorded ion current as function of the laser frequency and the number of segment illumination within a coaxial setup.

e) Suggestive estimate of coaxial sensitivity

In an additional preliminary experiment the sensitivity of a coaxial APLI approach within a stainless steel tube ($\varnothing_{inner} = 9$ mm, $l = 22$ cm; conically shaped end with 0.8 mm orifice) was investigated. For this purpose anthracene was deposited in a cryo-trap and the saturation vapor was balanced with nitrogen to a total pressure of 3000 mbar at constant temperature. A flow controller allowed for an accurate delivery of the compound through a small glass pipette sheathed by the main carrier gas. The final concentration in the ionization stage was calculated from the known vapor pressure at the prevailing room temperature, the mixing ratio after pressurizing with nitrogen and the mixing ratio within the total flow. The 248 nm laser beam was reduced to a circular shape of 1.5 mm in diameter and directed centrically along the tube. The pulse energy was 60µJ inside the tube with a repetition rate of 100 Hz. Considering an average drift velocity of $v_x = 0.4$ m·s^{-1} ($Q_{choked} = 1.4$ L·min^{-1}) a hypothetical segment was illuminated 55 times prior to entering the transfer capillary. In this way a standard addition method as shown in Figure 19 (left) was performed. The extrapolated linear regression revealed a background anthracene concentration of 3 pptV (parts per trillion by volume) prevailing in the used tube system (cf. Figure 19, right). It follows that even with considerably smaller laser beam dimensions and pulse energies condensed aromatic hydrocarbon compounds are detectable in the pptV range.

4 Results and Discussion

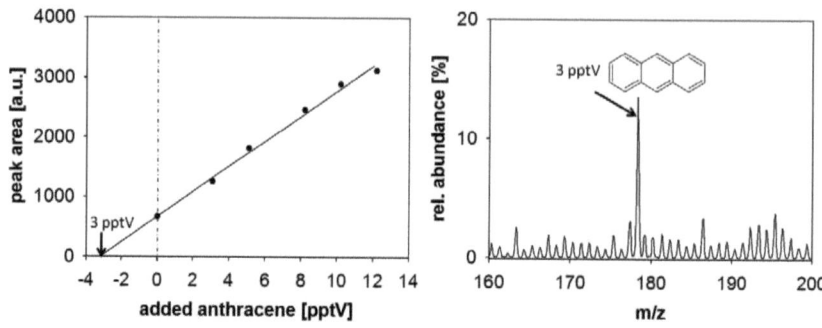

Figure 19: (left) Standard addition method for anthracene in a coaxial APLI configuration. (right) Coaxial APLI mass spectrum with 3 pptV anthracene present.

In summary the preliminary experiments showed: (i) The choked flow of the pressure reduction stage (transfer capillary) of a MS can be used to keep up a constant laminar flow within ionization tubes of inner diameters ranging between $Ø_{inner}$ = 9 mm and 4 mm, (ii) there is efficient ion transport over considerable distances within laminar flows at atmospheric pressure, (iii) metal tube systems in contrast to quartz/glass assemblies, are not exhibiting charge effects but show identical ion transmission behavior, (iv) the exact coaxial illumination along a laminar flow ionization stage significantly increases the ionization volume in contrast to conventional APLI, (v) the in-source detection efficiency of ions in such a setup is near unity, only diminished by diffusion losses, which also stands in contrast to conventional APLI, (vi) multiple illumination of a hypothetical infinitesimal segment, propagating with the average drift velocity v_x along the ionization tube, significantly increases the ion formation efficiency and (vii) a coaxial APLI assembly allows much smaller dimensioned laser beams and thus considerably smaller laser setups with sensitivities of AH still in the pptV range.

4.2.2 LFIS - Realization

Based on the preliminary experimental results a prototype of a laminar-flow ion source (cf. Figure 8) was constructed. Maintaining the laminar characteristic throughout the entire source up to the MS inlet was of primary importance. Therefore two critical parameters had to be resolved. First the bulk gas inlet into the ionization stage and second the transfer unit into

4 Results and Discussion

the capillary. The latter, as already indicated in the preliminary experiments, had to be of very sharp conical shape. For technical reasons, the utmost realizable full angle of 8°, ending in an aperture of 0.8 mm, was applied (cf. Figure 8). In order to prevent the propagation of any turbulence into the ionization stage special care had to be taken of the bulk gas inlet unit; however, technical reasons limited the feasible angle to 10° here (cf. Figure 8). The fairly wide inner diameter of $Ø_{inner} = 9$ mm ensures a gas inlet with an average drift velocity of 0.4 m·s^{-1}. A subsequent neck reduces $Ø_{inner}$ to 4 mm resulting in an average drift velocity of $v_x = 1.9$ m·s^{-1} within the ionization tube.

This design was first introduced in [48] and in cooperation with Bruker Daltonics GmbH, Bremen, a patent was filed in 2010 [79,80]. A more detailed technical description is given in section *3.1.3 Novel Laminar-Flow Ion Source (LFIS)*. A thorough fluid dynamic discussion based on computational modeling, the eventual APLI application as used for the intended atmospheric degradation product studies, and the VUV irradiation within the ionization stage will be given in the following three sections.

4.2.2.1 LFIS - Fluid Dynamical Simulations

a) Flow characteristic

Does the entire LFIS design meet the demand of maintaining the laminar flow? Does the realized inlet unit really provide the required smooth flow characteristics into the ionization tube or do generated turbulences propagate down to the cone? To shed more light into these issues fluid dynamical computational modeling, as for the small tube system in the preliminary experiments, were also performed for the designed prototype. The boundary conditions (dimensions and volume flow) were mapped one-to-one as listed in chapter *3.1.3 Novel Laminar-Flow Ion Source (LFIS)*. The computational results are shown in Figure 20. The solid lines in red represent stream lines, the size and direction of arrows indicate the velocity components and quantitative information of the prevailing velocity is obtained by the colored scale. Figure 20 (a) depicts the sample flow entering the inlet tube with the expected average velocity of around 0.4 m·s^{-1}. Albeit a soft "bouncing" effect on the bottom of the laser radiation inlet tube, indicated by the lower, thicker stream line, this simulation clearly demonstrates the smooth segue into an axial stream. No ablation from the main stream and no turbulences occur in that region. The only drawback obviously present is the appearance of a fairly high dead water volume (~3 cm^3) between the radiation inlet and the intersection of the

4 Results and Discussion

incoming gas flow. This is assumed to be negligible in terms of memory effects, but nevertheless leaves room for more improvement in further LFIS designs. Following the axial stream, the subsequent neck, which reduces the tubing diameter from 9 mm down to 4 mm is shown in a close up in Figure 20 (b). According to Venturi´s law the velocity increases, however, no significant dead water zone formation or laminar separation is visible at this point as well. The subsequent, fully developed laminar profile within the ionization tube becomes visible in Figure 20 (d). The color coded scale clearly reveals the expected Hagen-Poiseuille flow profile with a peak velocity of 3.5 m·s^{-1} and nearly 0 m·s^{-1} on the tube walls. Figure 20 (c) depicts a close-up of the fluid dynamic behavior within the conically shaped transfer stage to the MS transfer capillary. Again, no dead water zones or turbulences are formed, ensuring a smooth transfer into the MS. The velocity increases at this point up to 80 m·s^{-1} (exceeding the shown color scale; deep red), following again Venturi´s law.

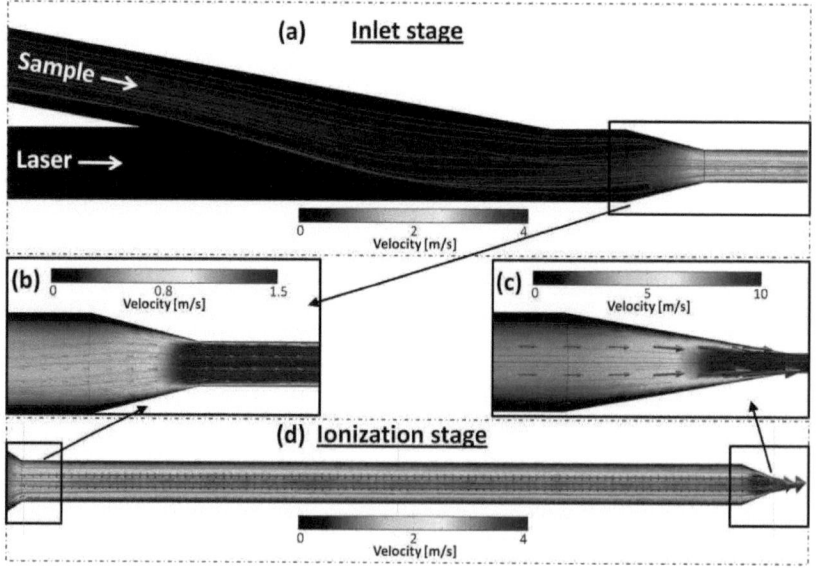

Figure 20: Fluid dynamical simulation of the LFIS: (a) Inlet stage, (b) close up of the neck unit, (c) close up of the cone and (d) ionization stage. Boundary conditions were used as described in chapter *3.1.3 Novel Laminar-Flow Ion Source (LFIS)*.

4 Results and Discussion

In conclusion, the fluid dynamical simulations fully support the present design of the LFIS in terms of maintaining a laminar flow throughout the entire setup. In contrast to the conventional API source (cf. Figure 14) the LFIS allows for rather accurate determinations of parameters such as dwell times, for predictions and for fairly precise calculations due to the well-characterized flow. The number of poorly characterized, uncontrollable and unstable source parameters, which are present in the commercially available API setup (cf. *4.1 Common* API *sources*) are reduced to the knowledge of the pressure within the tube system, the volume flow (determined by the transfer capillary) and the LFIS dimensions.

b) Diffusion along the flow propagation

The Comsol Multiphysics (Version 4.0a) software package is capable of modeling the spatial evolution of an initially confined analyte distribution. This approach gives valuable insight into the prevailing mechanisms of the LFIS and allows for qualitative as well as quantitative predictions concerning coaxial illumination, diffusion behavior and chromatographic peak broadening. Furthermore, this type of simulation enables the numerical treatment of possible ion chemistry, since the exact evolution of concentration and dwell times can be derived. The purpose of the present simulation, however, is to merely give an impression of the general evolution of a confined volume segment containing the analyte and to derive some qualitative estimates. The boundary conditions for the general fluid dynamical simulation were the same as described before; additionally a volume segment containing the analyte with a diameter of 0.5 mm and a concentration of ~5 ppbV was placed on an axial position of the ionization tube and the diffusion coefficient of toluene was used ($6.98 \cdot 10^{-2}$ $cm^2 \cdot s^{-1}$ [100]). Figure 21 shows a selection of frames obtained within the simulation run. The most apparent result is the rapid decrease of the maximum concentration by an order of magnitude within the first 5 ms (10^{-11} $mol \cdot cm^{-3}$ to 10^{-12} $mol \cdot cm^{-3}$), also illustrated in the plot on the right. The exponential decay continues down to 10^{-14} $mol \cdot cm^{-3}$ after 70 ms. This demonstrates an important consequence concerning the multiple illumination of a hypothetical segment with a small laser beam diameter. Considering a pulse repetition rate of 200 Hz one segment would be illuminated every 5 ms. Thus, between every laser shot the concentration of the formerly generated ions is decreased by at least a factor of ten, effectively circumventing localized ion saturation effects within the ionization volume. Furthermore, rough estimates concerning chromatographic peak broadening (note, that no LC

4 Results and Discussion

or GC is applied within this work) can be derived from the maximum spatial broadening. Assuming a ~5 cm spatial broadening that propagates with an average axial velocity of $v_x = 1.9$ m·s^{-1} gives a temporal broadening of ~26 ms, which is far from being the limiting factor in common chromatographic peak broadening. This simulation further illustrates the typical Hagen-Poiseuille flow profile in which the initial spot is dispersed with downstream propagation.

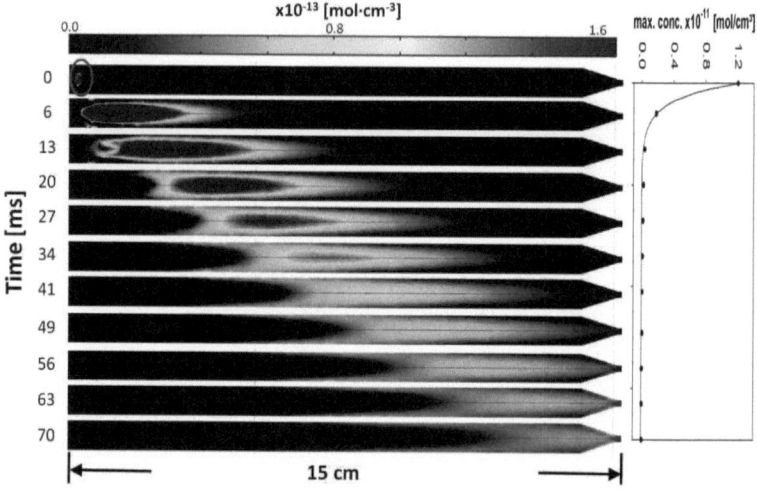

Figure 21: (left) Simulated spatial evolution of an initially localized ion packet along the downstream propagation of the LFIS and (right) corresponding plot of the time dependent evolution of the maximum concentration.

4.2.2.2 LFIS – APLI

a) Choice of laser system

The preliminary experimental section demonstrated the excellent attainable sensitivity via coaxial illumination with significant smaller laser beam diameters and lower pulse energies than applied in conventional APLI. In general, decreasing the energy output allows for the use of considerably smaller and thus more convenient APLI-MS setups. As mentioned by Short et al. [65], "APLI is not yet widely used due to the large size, expense and maintenance associated with the laser system". Accordingly, a compact diode pumped solid state laser was purchased in cooperation with Bruker Daltonics GmbH, Bremen. This type of

4 Results and Discussion

laser provides beam diameters of $d_{beam} = 0.5$ mm and pulse energies of $E_{pulse} = 60$ µJ with $t_{pulse} = 1$ ns pulse duration, accounting for power densities of $3 \cdot 10^7$ W·cm^{-2}. The DPSS significantly outperforms the commonly used exciplex laser in terms of size (cf. Figure 6), acoustic noise level (DPSS: around 0 dB; exciplex: 70 dB [101]), maintenance and purchase cost (DPSS: ~ 10 k€; Exciplex: ~ 40 k€) [44]. Consequently, the implementation of such a laser system generally renders the technique of APLI-MS more practicable and attractive. At this point it is stressed that comparative experiments between the performance of a DPSS and an exciplex laser in the conventional perpendicular APLI-MS setup were carried out [44]. Here, a 0.01 µM pyrene solution in methanol was delivered in a continuous flow mode and both laser beams were manually signal optimized. The recorded mass spectra revealed a relative signal intensity of 0.9 % of the DPSS laser generated signal as compared to the signal obtained with the exciplex laser. The need for significant increase in ionization volume along the light propagation direction was pinpointed, when the use of smaller laser systems but comparable performance as currently obtained with exciplex lasers, is desired. Consequently, decreasing the size of the laser system in APLI-MS is directly related to the new laminar flow design.

b) Laser beam expansion.

The power density output of the DPSS laser exceeds the exciplex laser by an order of magnitude when no further optical devices are used (cf. Table 1). In the first instance this leads to the assumption of beam expansion and hence an increase of the ionization volume $V_{ionization}$ [cm^3] according to

$$V_{ionization} = \left(\frac{d_{beam}}{2}\right)^2 \cdot \pi \cdot l_{ionization\ tube} \qquad (eq\ 8)$$

with d_{beam} = beam diameter [cm] and $l_{ionization\ tube}$ = length of the ionization tube [cm]. On the other hand, such an approach proportionally decreases the photon flux Φ [photon s^{-1} cm^{-2}] with the exact same quadratic dependency on the diameter according to

$$\Phi = \frac{E_{pulse}}{t_{pulse} \cdot \left(\frac{d_{beam}}{2}\right)^2 \cdot \pi \cdot h \cdot \frac{c}{\lambda}} \qquad (eq\ 9)$$

with E_{pulse} = pulse energy [J], t_{pulse} = pulse duration [s], h = Planck´s constant (6.626·10^{-34} J·s), c = velocity of light (3.0·10^8 m·s^{-1}) and λ = wavelength [m]. In order to evaluate a reasonable

4 Results and Discussion

implementation of a beam expending unit, prior to coaxial illumination, the following theoretical considerations were made: The total number of ions $n_{[M]^+}$ generated within one laser pulse and transferred into the MS is described by

$$n_{[M]^+} = \{[M]^+\} \cdot V_{ionization} \qquad (eq\ 10)$$

with $\{[M]^+\}$ = the radical cation concentration [molecules·cm^{-3}]. Furthermore, the generation of $\{[M]^+\}$ within one laser pulse is in general obtained from

$$\{[M]^+\} = \{[M]\} - \{[M]\} \cdot e^{-k_{ion}t_{pulse}} \qquad (eq\ 11)$$

with $\{[M]\}$ = neutral analyte concentration within the ionization tube [molecules·cm^{-3}] and k_{ion} = first order rate constant of ion formation [s^{-1}]. However, the change of $\{[M]\}$ within one laser pulse is negligible, and equation 11 can approximately be reduced to

$$\{[M]^+\} \approx \{[M]\} \cdot k_{ion} \cdot t_{pulse} \qquad (eq\ 12).$$

For an ideal two-photon absorption process the rate constant of ion formation k_{ion} is obtained from equation 13

$$k_{ion} = \sigma \cdot \Phi^2 \cdot \varphi_1 \cdot \varphi_2 \qquad (eq\ 13)$$

(σ = absorption cross section of the overall (1+1) REMPI process [cm^4·s·molecule^{-2}]; $\varphi_{1,2}$ = quantum yield for each step of the two absorption processes [molecule·photon^{-1}]). Equations 13 and 12 show the quadratic dependency of the ion concentration $\{[M]^+\}$ on the photon flux. Taking into account the reciprocal quadratic dependency of the photon flux Φ on the beam diameter d_{beam}, as shown in equation 9, the following proportional relation between the ion concentration $\{[M]^+\}$ and the beam diameter after one laser pulse is obtained:

$$\{[M]^+\} \sim \frac{1}{d_{beam}^4} \qquad (eq\ 14).$$

Combining the quadratic dependency of $V_{ionization}$ on the beam diameter (equation 8) with equations 10 and 14 leads to the following proportional relation for the number of ions generated after one laser pulse:

$$n_{[M]^+} \sim \frac{1}{d_{beam}^4} \cdot d_{beam}^2 \quad \rightarrow \quad n_{[M]^+} \sim \frac{1}{d_{beam}^2} \qquad (eq\ 15)$$

4 Results and Discussion

Consequently, doubling the beam diameter would result in a four-fold ionization volume and ideally in a four-fold higher ion yield $n_{[M+]}$, but according to equation 15 the influence of the lower photon flux concurrently diminishes the ion yield by a factor of 16. Hence, expanding the laser beam for an ideal two-photon absorption process is not reasonable. In some cases, however, it was shown that the quadratic photon flux dependency of the ion formation is valid only for low power densities. Zakheim et al. [102] have reported on a detailed kinetic model describing feasible stimulated emission concurrently occurring at higher photon fluxes and thus the first electronically excited state starts being kinetically saturated. In case of kinetic saturation equation 13 has to be rewritten

$$k_{ion} = \sigma \cdot \Phi^b \cdot \varphi_1{}^b \cdot \varphi_2 \quad \text{(eq 16)}$$

with b = experimentally determined number. Consequently, the following expression for the proportionality between the number of ions $n_{[M+]}$ and the beam diameter d_{beam} is then obtained

$$n_{[M]^+} \sim \frac{1}{d_{beam}{}^{2 \cdot b}} \cdot d_{beam}{}^2 \quad \text{(eq 17)}.$$

Apparently, the reasonable implementation of a beam expanding unit for the coaxial APLI depends on the exponential term b of the photon flux. In case of $b = 2$ the above mentioned ideal two-photon absorption process is obtained. For $1 < b < 2$ stimulated emission and radiationless transition processes depleting the electronically excited state come into effect (cf. Figure 2). Hence, increasing the photon flux at this point concurrently induces the depletion of the excited molecule population. According to equation 17, an increase in d_{beam} still results in ion loss. With $b = 1$ the number of generated ions becomes independent of the beam diameter. At this point the resonant state is completely saturated; depletion processes from the excited to lower states and absorption from the ground to the excited state are in dynamic equilibrium. With $0 \leq b < 1$ the number of generated ions is proportionally dependent on d_{beam} and hence increasing the ionization volume becomes favorable. The spectroscopic kinetics occurring in this range of b is characterized by saturation of the transition from the excited state into the ionization continuum. Typically, the constant b is obtained from a doubly logarithmic plot of the ion signal as function of the power density. Combination of equations 16 and 12 leads to the following linear form

$$log\{[M]^+\} = b \cdot log\Phi + log(\sigma \cdot \varphi_1{}^b \cdot \varphi_2 \cdot t_{pulse}) \quad \text{(eq 18)}$$

4 Results and Discussion

and b is simply derived from the slope of the curve. Figure 22 depicts two such plots obtained for the DPSS laser operated in (i) the conventional perpendicular (left) and (ii) a modified coaxial configuration (right), respectively.

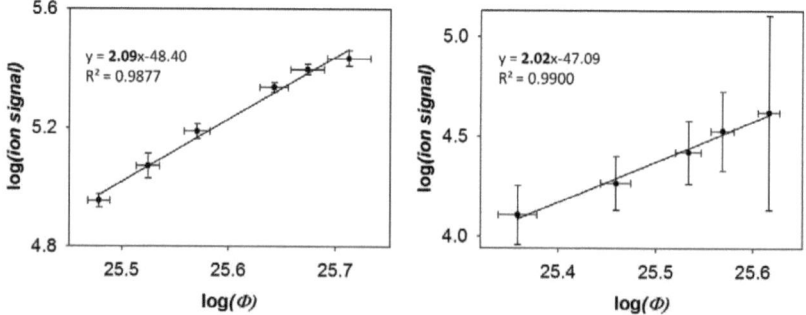

Figure 22: Power plot of the ion signal versus the photon flux Φ of the DPSS laser. (left) In the conventional perpendicular configuration with a 0.01 µM pyrene solution in methanol present, added in a continuous flow mode, and (right) in the modified coaxial LFIS setup with ~ 6 ppbV pyrene vapor present.

For the first setup a pyrene solution in methanol (0.01 µM) was added in a continuous flow mode into the ion source (cf. Figure 7). In (ii) a shortened LFIS version ($l = 5$ cm) was connected to the capillary with solid pyrene deposited on the tube walls, thus providing an estimated analyte concentration of 6 ppbV (assuming an equilibrated state with the bulk gas – cf. 3.1.3 *Setup for Neutral Radical Induced ITP Studies [82]*). The power density was varied by simply changing the energy output via the laser software setting between 45 µJ and 77 µJ per pulse (at 200 Hz) and 35 and 60 µJ (at 10 Hz), respectively. Linear regression according to equation 18 resulted in (i) $b = 2.09$ and (ii) $b = 2.02$. The error bars for the photon flux were derived from an estimated variation in pulse energy of $E_{pulse} = \pm 1.5$ µJ and for the ion signal the standard variation of the averaged MS signal of a two minute record was used. For a more quantitative evaluation of b these experiments should again be performed under more precise conditions and with a power density range of at least one order of magnitude. Nevertheless, the results clearly demonstrate that the exponential term of the photon flux is well above $b > 1$, which means that in both cases the photon flux, as provided by the DPSS laser (~10^{25} photons·cm^{-2}·s^{-1}), induces a nearly ideal two-photon process with no observable

4 Results and Discussion

spectroscopic saturation effects. Consequently, according to equation 17, doubling the beam diameter would result in a nearly quadratic decrease in the number of ions $n_{[M]+}$ generated after one laser pulse. Besides these theoretical considerations, experimental losses due to laser beam expansion have to be considered as well. These encompass pulse energy losses evoked by the additionally needed optical devices, and enhanced diffusional losses of already generated ions. The latter is due to an increased initial spatial distribution of generated ions, which eventually reaches the tube dimensions in an earlier point of time along the downstream propagation of the gas flow (cf. Figure 21).

In conclusion, the implementation of a beam expansion unit prior to coaxial illumination with the DPSS laser was demonstrated to be of minor importance, if at all.

c) Interaction laser radiation → metal surface

The approach of coaxial illumination in the LFIS is inevitably accompanied by the interaction of the metal surface and the laser radiation. Even with the careful exact axial positioning, the beam will eventually strike the cone (cf. Figure 8) and the most obvious result is the formation of photoelectrons. NIF processes could definitely be enhanced (cf. *1.2.1.3 Negative Ion Formation (NIF)*), however, the positive MS mode would be adversely affected. Novotny et al. [103] have measured bimolecular recombination rates of some polycyclic aromatic hydrocarbon cations with electrons in the order of 10^{-6} cm^3·s^{-1}. Consequently, it seemed necessary to provide an inner surface reducing the emission of photo electrons. Thus the entire tube system was nickel-plated since the used radiation at 266 nm corresponds to 4.66 eV and the work function of nickel is around 5 eV [104]. However, this approach merely affects the single photon induced emission of electrons: electron emission from multiple photon absorption processes, as described by Logothetis et al. [105], may still occur. Furthermore, in 2010 Brockmann et al. [106] introduced time resolved current measurements that were performed with the setup for capillary investigations (cf. *3.1.4.1 Setup for Characterization of Transfer Capillaries*). Herein the LFIS was connected to the transfer capillary with a coaxially directed laser beam. The measured current at the detection sieves was recorded temporally resolved for each laser pulse. As shown in Figure 23 the recorded signal shows an unexpected high rise shortly after the laser pulse, indicating non proportionally elevated ion formation within the cone and the small transfer region into the capillary. The subsequent decay of the slope is shaped as expected and is in full accordance

4 Results and Discussion

with the diffusional losses along the tube as shown in the perpendicular experiment in Figure 17.

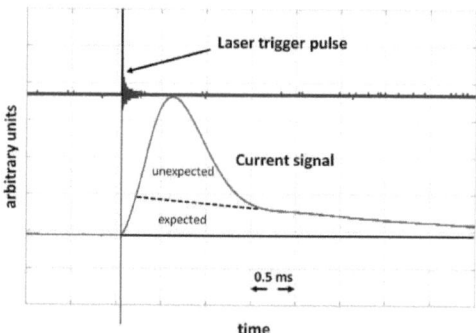

Figure 23: Time resolved current measurements using the laminar flow ion source [106]. The lower curve represents the evolution of the current that is induced by an ion packet, which was generated by one coaxial laser pulse.

A first tentative explanation for the unexpected behavior is partially based on VUV contributions via second-harmonic generation of the initial 266 nm radiation on the metal surface of the cone. Such effects are well known and for example described by Chen et al. [107]. It follows that single photon ionization absorption cross sections of the analyte additionally have to be considered, which are fairly large and thus might significantly increase the partial ionization efficiency within the cone. Both effects, the multiple photon induced photo electron emission and the second-harmonic generation are competitive processes and have been investigated for example by Tomas et al. [108]. However, the experimental conditions in the literature are not comparable to the present LFIS setup, rendering meaningful conclusions nearly impossible. Thus, the impact of the laser radiation-/ metal surface interaction on APLI-MS within the laminar-flow ion source will have to be subject to further investigations.

4.2.2.3 LFIS – APPI

a) Implementation

Figure 8 shows the implementation of an APPI unit between the cone and the ionization tube. A schematic of the segment itself is shown in Figure 24. In the first instance,

4 Results and Discussion

special attention was focused on conserving the laminar flow conditions by cutting a flute into the LiF window, which exactly matched the inner tubing dimensions and thus provided a smooth inner surface. However, machining an optical device requires sophisticated machinery and skills to maintain a sufficient optical transmission of the treated surfaces. A small titanium hardened drill was used to grind the flute in several layers at high drill rotation speed. The machined window is mounted o-ring sealed; thus the entire LFIS can still be pumped down to 10 mbar to provide convenient repeated flushing sequences of the entire system. The commercially available APPI lamp as used in the conventional API setup as well as home-built spark discharge lamps can be used for VUV radiation supply. Since the inner tubing diameter is 4 mm, sufficient irradiation intensity is provided throughout the entire ionization volume with respect to absorbing matrix components (cf. *1.2.1.2 Atmospheric Pressure Photo Ionization (APPI)*).

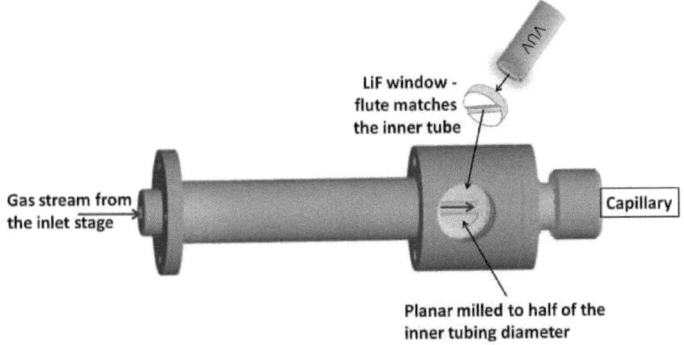

Figure 24: Schematic of the APPI unit of the LFIS.

b) Transit time.

The above time resolved current measurements have also been performed with laser radiation directed through the LiF window of the APPI inlet stage, i.e. perpendicular to the gas flow. The delay between the trigger pulse and the fastest ions arriving at the detector was measured to be 5 ms. Thus, ions generated with VUV radiation at the designated position are transferred downstream to the exit of the capillary and reach nearly collision free conditions within 5 ms (the transfer time from the first differential pumping stage into the high vacuum region –cf. Figure 4, right- is assumed to be negligible).

4 Results and Discussion

4.2.3 Development of APPI on Transfer Capillaries

Several degradation studies were mass spectrometrically monitored, using the LFIS with the APPI unit as ionization source. It soon became apparent that the determined dwell time of 5 ms (cf. *4.2.2.3 LFIS – APPI*) resulted in sufficient collisions to significantly impact on the ion distribution of a degradation product study up to the point of complete loss of reasonable mass spectrometric information. A discussion on the prevailing ITPs will be given in chapter *4.3 Ion Transformation Processes*. Here, merely the demand of notably reducing the total number of collisions between the ionization and the detection step is stressed. Assuming the gas kinetic collision rate for standard conditions (at 298 K), around 10^6 collisions occur for every ms of dwell time, hence, lowering the pressure within the ionization region and/or decreasing the time between VUV irradiation interaction and entering the collision free region was required. Consequently, moving the VUV ionization position further downstream was the only reasonable approach. As a first step, a VUV lamp was built into the first differential pumping stage located between the transfer capillary exit and the MS sampling skimmer (cf. Figure 4). However, the realized setup with a home-built spark discharge lamp as radiation source (cf. *4.2.3.3 Development of Miniature VUV Spark Discharge Lamps*) resulted in rather poor sensitivity, owing mostly to the lower analyte density and more importantly due to a spatially fairly wide spread ion distribution. This assumption is supported by numerical ion trajectory simulation, as shown in Figure 25 (right).

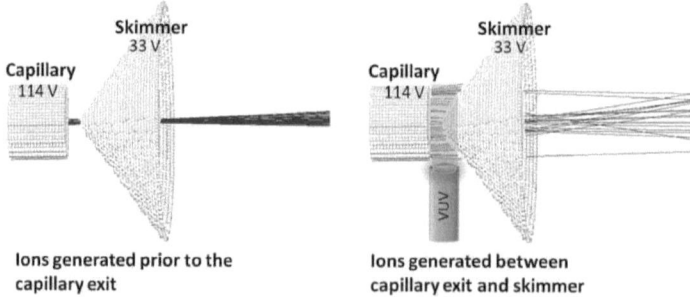

Ions generated prior to the capillary exit

Ions generated between capillary exit and skimmer

Figure 25: Results of an ion trajectory calculation for the transfer efficiency from the capillary through the skimmer with (left) ions being generated upstream of the capillary exit and (right) ions being generated between capillary exit and skimmer.

4 Results and Discussion

The latter obviously affected the sampling efficiency of the skimmer into the high vacuum region, as compared to the ion containing gas stream exiting the capillary (cf. Figure 25, left). Hence, this setup was not further considered. The feasible second approach was the installation of a VUV radiation source directly onto the transfer capillary. This would (i) preserve the transmission efficiency through the MS skimmer and (ii) lead to ionization at appreciable neutral analyte densities. Since the transfer capillary is completely opaque for radiation below 300 nm the challenge was to get optical access to the capillary gas stream. This approach called for an exploration of completely new research areas, which will be the subject of the following chapters. A detailed experimental and theoretical study concerning the fluid dynamical behavior of the capillary, with additional determination of ion transmission efficiencies will be presented. In the course of confining the neutral analyte distribution to the inner dimensions of the capillary, home-built miniature VUV spark discharge lamps will be introduced as the more appropriate radiation source in contrast to the commercially available large dimensioned lamps. Additionally, an in-depth characterization of the spark discharge VUV lamp will be presented. Finally the impact of different ionization positions with respect to reducing ITPs will be demonstrated.

4.2.3.1 Characterization of Transfer Capillaries

a) Comparison of original and home–made capillary

Prior to invasive machining it was necessary to investigate and compare the flow characteristics and ion transmission efficiencies of the original transfer capillary (purchase cost: ~1000 €) with home-made capillaries from bulk stock (Hilgenberger GmbH, Malsfeld, Germany; purchase cost: ~10 €). Obviously, the purchase cost did not allow such experiments with the original capillaries, and significant inferior performance of the home-made capillary would have made the proposed ionization approach useless. Therefore, the setup as sketched in Figure 10 was constructed to carry out appropriate experiments. Figure 26 depicts two plots for performance comparison. On the left, the volume flow is plotted as function of the background pressure in the recipient (p_1), with 970 mbar at 298 K on the high pressure port (p_0). The dimensions of both capillaries were $l = 18$ cm and $Ø_{inner} = 0.6$ mm (as stated by the manufacturers; with ± 0.01 mm tolerance for the inner diameter of the home-made capillary). Both curves overlap almost perfectly, demonstrating no significant difference in terms of the flow dynamical behavior. The obtained choked flow of the original capillary is slightly higher

4 Results and Discussion

than that of the home-made one (Q_{choked} = 1.41 and 1.37 L·min^{-1}), indicating small differences in the inner diameter of about 0.007 mm (see below).

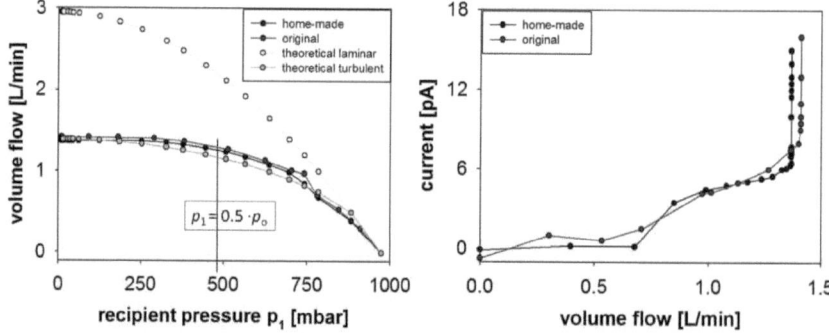

Figure 26: (left) Flow dynamical characterization of a home-made and an original transfer capillary. The volume flow is plotted as function of the varied pressure on the downstream side of the capillary, with an upstream stagnation pressure of 970 mbar. (right) Comparing ion transmission measurements of a home-made and an original transfer capillary.

In Figure 26 (right) a comparison of ion transmission curves is shown. Here a 248 nm laser beam was directed coaxially into the LFIS with pyrene as the analyte as described before. The beam was reduced to 6 mm in diameter and a pulse energy of 0.7 mJ at 100 Hz repetition rate was applied. The current recorded at the detector sieve was plotted against the volume flow. Again, both curves overlap nicely. The rampant trend in both curves, once the choked flow is established (note that the pressure p_1 is still further reduced) has not been clarified yet. This trend is tentatively assigned to ion mobility effects in the transfer region from the end of the capillary to the Faraday cup (cf. Figure 10), thus more likely an effect of the setup than of any fluid dynamic property within the capillary. Nevertheless, this shape was of high reproducibility and hence rendered the comparison of the relative ion transmission efficiencies possible. Furthermore, investigations on the absolute ion transmission efficiency of a capillary revealed exceptionally high transmission factors of up to 0.5 relative to base current measurements without a transfer capillary [106]. This result is in good agreement with the observation made by Lin and Sunner [109], who measured the current transmitted through glass, metal and Teflon capillaries with lengths of up to 15 m and inner diameters ranging

4 Results and Discussion

between 1 and 4 mm. However, there is hardly any literature available on the fluid dynamical and ion transmission behavior of MS transfer capillaries leaving much room for further investigations. Conclusively, the properties of the original and a home-made transfer capillary in terms of flow dynamics and ion transmission were demonstrated to be identical. This result paved the way for developing the novel APPI approach and it initiated the extended production of home-made capillaries in a great variety (chemically metalized end caps, metalized inner surfaces, flared entrance, etc.) of which many have been successfully applied in several types of mass spectrometers.

b) Adaptability of fluid dynamic equations - laminar or turbulent

As mentioned above, Figure 26 (left) depicts the flow behavior with varying differences between the stagnation pressure p_0 and recipient pressure p_1. From theory (e.g., [99]) it is known that the choking of a flow inside a tube should occur at around $p_1 = 0.5 \cdot p_0$. This is in good accordance with the observations made here. It is noted that in case of any turbulences occurring close the segue into the capillary the actual reference pressure p_0 is lower. The impact of the sharp cone of the LFIS on the effective stagnation pressure is shown later. Figure 26 (left) furthermore shows two calculated curves, which are based on empirical fluid dynamic equations [99,110], developed for describing the flow behavior of large tube systems for the laminar

$$Q_{laminar} = \frac{8100 \cdot \left(\frac{\emptyset_{inner}^4}{l} \cdot \frac{p_0^2 - p_1^2}{2} \right)}{p_0} \quad \text{(eq 19)}$$

and the turbulent case, respectively:

$$Q_{turbulent} = \frac{8040 \cdot \emptyset_{inner} \cdot \left(\frac{\emptyset_{inner}^3}{l} \cdot \frac{p_0^2 - p_1^2}{2} \right)^{\frac{4}{7}}}{p_0} \quad \text{(eq 20)}.$$

Here $Q_{laminar/turbulent}$ denotes the volume flow in [L min^{-1}], $\emptyset_{inner}$ is the inner diameter in [cm] and l describes the length of the tube in [cm]. Apparently, the shape of the calculated turbulent case, with $\emptyset_{inner} = 0.0609$ cm and $l = 18$ cm overlaps very well with the experimentally determined shape. Two important conclusions are drawn from this result: (i) Well known empirical fluid dynamical equations are to a good approximation valid for the dimensions of the present MS transfer capillaries and (ii) the gas flow through a transfer

4 Results and Discussion

capillary, with operating conditions typically prevailing in the experimental setup (gas temperature ≈ room temperature = constant; upstream stagnation pressure $p_0 \approx 1000$ mbar; first MS pumping stage pressure of $p_1 \approx 4$ mbar), is characterized by a fully developed turbulent choked flow ranging between 0.8 and 1.4 l/min (for $\varnothing_{inner} = 0.05 - 0.06$ cm). According to equation 5 the turbulent characteristic is further substantiated by the calculated Reynolds number of $Re = 3310$, assuming an average drift velocity of $v_x = 83$ m·s^{-1} and an inner diameter of $\varnothing_{inner} = 6 \cdot 10^{-4}$ m. This is well above the critical number of $Re_{crit} = 2300$ [99], consequently, essentially turbulent flows prevail in these capillary systems under typical operating conditions. Equation 20 furthermore offers a valuable tool for determining the inner capillary diameter. With the experimentally established parameters (l, p_0, p_1 and Q_{choked}) $\varnothing_{inner}$ can be fitted within ± 1 µm accuracy. In this way the inner diameters of the original and the home-made capillary were calculated to be $\varnothing_{inner} = 0.612$ mm and $\varnothing_{inner} = 0.605$ mm, respectively. The latter is in full accordance with the manufacturer tolerance declaration of 0.6 ± 0.01 mm.

c) Critical and static pressure, velocity distribution, and transit times

The experimental validation of the available mathematical descriptions for turbulent flow inside capillaries allowed the calculation of parameters such as the minimum or critical pressure, the static pressure, the velocity distribution *along* the main capillary axis and eventually the transit times from a certain location inside the capillary to the exit port. The following experimental data were required: (i) The length l [cm] of the capillary, (ii) the choked flow Q_{choked} [L·min^{-1}], (iii) the pressure p_1 [mbar], and (iv) the upstream stagnation pressure p_0 [mbar]. As mentioned above, the latter is fairly critical, since turbulences around the entrance of the capillary might result in an initial significant pressure drop. Pressure measurements at some distance to the capillary entrance might not properly reflect p_0. However, due to the conically shaped coupling stage of the laminar-flow ion source such turbulences are very well minimized, as has been shown elsewhere [48], so that to a good approximation the measured pressure (cf. Figure 8) does correctly reflect p_0. This fundamental information is first used to fit the inner diameter $\varnothing_{inner}$ according to equation 20, until the calculated and the observed choked flow match (see above). Subsequently, the minimum or critical pressure p_{crit} [mbar], where the gas velocity reaches sonic speed, can be calculated as follows [99]:

4 Results and Discussion

$$p_{crit} = \frac{4.51 \cdot \left(\frac{\emptyset_{inner}^3 \cdot p_o^2}{2 \cdot l}\right)^{\frac{4}{7}}}{\emptyset_{inner}} \quad \text{(eq 21)}.$$

For the dimensions of the capillaries (for $\emptyset_{inner}$ = 0.05 – 0.06 cm) and the operating conditions used in the mass spectrometer ($p_o \approx$ 1000 mbar and $p_1 \approx$ 4 mbar), the critical pressure ranges between p_{crit} = 180 - 220 mbar. The gas stream exits the capillary with p_{crit} into the background pressure p_1 of the first differential pumping stage, with $p_{crit} \approx 50 \cdot p_1$. The consequences with respect to the appearance of the resulting type of gas expansion are a topic of controversial discussion, if at all, and still leave much room for further research.

It follows that the static pressure $p_{static(x)}$ [mbar] evolution along the transfer capillary can be calculated according to [110]:

$$p_{static(x)} = p_0 \cdot \sqrt{1 - \left(1 - \frac{p_{crit}}{p_0}\right) \cdot \frac{x}{l}} \quad \text{(eq 22)}$$

with x [cm] as the position of the capillary, relative to the entrance. The velocity distribution $c_{(x)}$ is then obtained by [110]:

$$c_{(x)} = \frac{p_0}{p_{static(x)}} \cdot c_{(0)} \quad \text{(eq 23)}$$

with $c_{(0)}$ as the entrance velocity [m·s^{-1}], which is simply derived from dividing the flow Q_{choked} by the area cross-section of the capillary (assuming the *fitted* diameter). Eventually, the residence time t_{res} of an ion within the transfer capillary is derived from numerical integration over the fraction of $x/c_{(x)} = t_{res(x)}$ according to

$$t_{res} = \sum_{x=ionization\ positon}^{x=L} t_{res(x)} \quad \text{(eq 24)}$$

It is noted that equations 20 - 21 hold true for air as the bulk gas at 293 K, and equations 22 - 24 assume isothermic conditions, i.e., the gas temperature inside the capillary, despite the expansion process, to be constant. This very well coincides with the experimental conditions of a common degradation experiment when sampling from the reactor. It is noted here that in separate works the isothermic condition for the gas expansion inside the capillary was experimentally verified. In Figure 27 typical results of calculations based on the above equations are shown.

4 Results and Discussion

Table 2: Calculated dwell times for ions within the capillary as function of the ionization position.

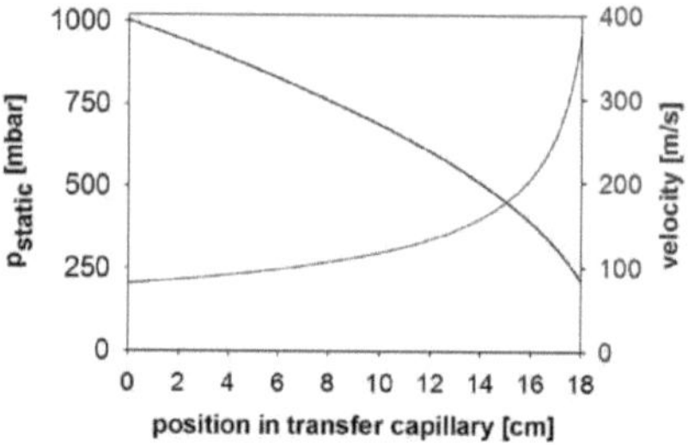

ionization position [cm]	static pressure [mbar]	dwell time [ms]
0.0	1003	1.55
3.0	920	1.19
6.0	828	0.87
9.0	725	0.58
12.0	605	0.33
15.0	453	0.13
17.0	314	0.03
17.5	268	0.02

Figure 27: Static pressure and velocity distribution within the transfer capillary, calculated for the following conditions: $\varnothing_{inner} = 0.61$ mm, $Q_{choked} = 1.41$ l·min^{-1}, $p_0 = 1003$ mbar, $p_1 = 3$ mbar;

At this point it is worth mentioning that the static pressure inside the capillary became measurable with opening up the capillaries (cf. Figure 9 and chapter *4.2.3.3 Development of Miniature VUV Spark Discharge Lamps*). At several positions holes were drilled into the capillary body using different types of capillaries ($\varnothing_{inner} = 0.5 - 0.6$ mm). In all cases the measured static pressure and the calculated pressure was within ±2 % accuracy. Additional time resolved current measurements, as introduced in chapter *4.2.2.3 LFIS – APPI*, revealed dwell times of around 1.3 ms for ions being generated directly at the entrance of the transfer capillary, essentially at position $x = 0$ (cf. Table 2). These results were further strong experimental support for the applied equations.

The closest ionization position within the laminar-flow ion source (cf. *4.2.2.3 LFIS – APPI*) revealed a minimum transit time of 5 ms and it was shown earlier that minimum transit times of 5 ms are also obtained with commercially available API sources [46]. Compared to the results in Table 2 the approach of ionizing within the capillary is thus capable of reducing residence times in the high collision rate region by a factor of 250. The ionization efficiency is expected to be fairly high since the analyte density is just reduced by a factor of four.

d) Upstream pressure variation

It is worthwhile mentioning that the stagnation of the choked flow is only valid for a constant upstream pressure p_0. Figure 28 (left) depicts the calculated impact of p_0 on the choked flow rate within a range of 400 mbar up to 2000 mbar, with $p_1 = 4$ mbar, and capillary

4 Results and Discussion

dimensions of $\emptyset_{inner} = 0.609$ mm and $l = 18$ cm. As can be seen, the choked flow varies within a range of around 0.3 L·min^{-1}. In the same manner, the critical pressure is plotted as a function of p_o in Figure 28 (right). A nearly linear behavior (according to equation 21: $p_{crit} \sim p_o^{8/7}$) is obtained, which means that variations in the upstream pressure significantly affect the properties of the gas stream entering the first differential pumping stage, as expected. Apparently, the stagnation of the choked flow and subsequently involved parameters, e.g., p_{crit}, p_{static} (cf. equation 22), holds true only for changes on the low pressure side, whereas varying p_0 does have significant impact. It follows that in terms of keeping the transmission properties of the subsequent ion optics (e.g., skimmer, multi poles, lenses - cf. Figure 4 right) constant, p_0 has to be treated with special attention. In addition, variations of the static pressure would particularly affect the windowless design of the implemented discharge lamps, as will be described in detail in chapter *4.2.3.3 Development of Miniature VUV Spark Discharge Lamps*.

When using the windowless spark discharge VUV lamps as ionization sources in the atmospheric chemistry degradation experiments, it was necessary to (i) implement a pressure gauge into the LFIS (cf. Figure 8), (ii) balance the pressure of the reactor (cf. *3.3.3 MS Sampling Unit*) and (iii) ensure a constant, smooth segue of the gas stream through the cone into the transfer capillary.

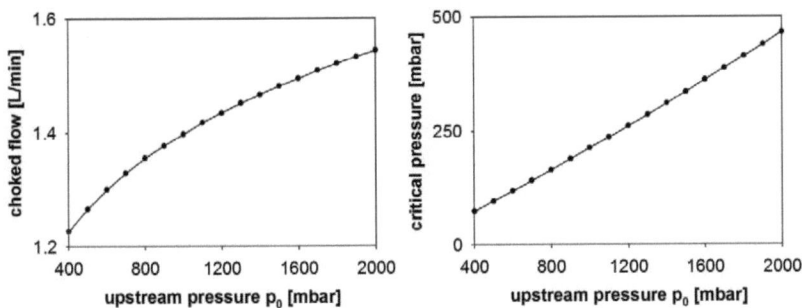

Figure 28: Calculated plots of (left) the choked flow and (right) the critical pressure as function of the upstream pressure. The low pressure side was assumed to be constant at 4 mbar, the capillary dimensions were $\emptyset_{inner} = 0.609$ mm and $l = 18$ cm.

4 Results and Discussion

4.2.3.2 First APPI on Capillary Approach

The first approach of irradiating the bulk gas in a capillary with light below 200 nm wavelength was similar to the APPI stage of the laminar-flow ion source (cf. *4.2.2.3 LFIS – APPI*). The capillary was grinded at 15 cm (cf. Table 2) to half its diameter within a length of one cm. It is worth mentioning that the first capillaries were manually machined; in the course of the experiments a computer controlled milling machine was used. According to the LFIS design a LiF window of one cm diameter and three mm thickness was modified with a flute exactly matching the surface of the inner tube. In several subsequent drying steps both pieces were glued to each other with an epoxy adhesive (UHU plus endfest 300, 2-K-Epoxidharzkleber). It is worth mentioning that despite using retail resin, no mass signals attributable to the adhesive have been observed. The long term impact of VUV radiation on the adhesive is not known, however, it is expected to turn brittle with time. In a first attempt the commercially available APPI lamp was used as radiation source and simply mounted onto the window. In this way an APPI stage within a capillary was designed, as shown in Figure 29.

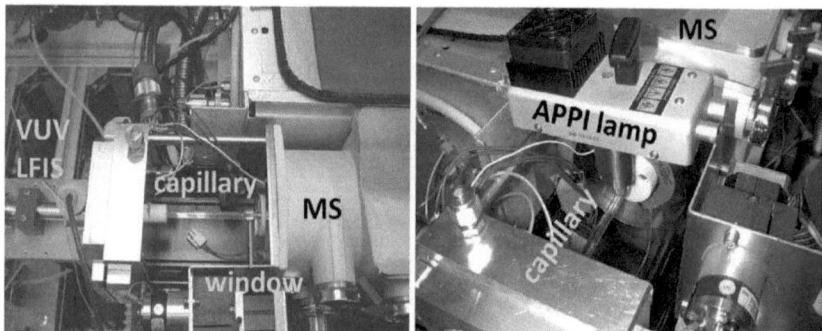

Figure 29: Photographs of (left) the LiF window mounted on the transfer capillary and (right) "APPI on transfer capillary" with the commercially available Kr-RF low pressure discharge lamp.

As envisioned, this approach gave reasonable results in terms of ITP reduction corresponding to the calculated dwell time of 0.13 ms (cf. Table 2 and *4.2.3.4 Impact of Different Ionization Positions on MS Spectra*). However, sensitivity was not as satisfying as expected, which was basically assigned to inefficient irradiation of the capillary gas flow. The APPI lamp with a

4 Results and Discussion

diameter of one cm illuminates a total area of 79 mm^2, however, the effectively illuminated gas flow area is maximal about 6 mm^2, hence a substantial part of the radiation is not available for ionization.

4.2.3.3 Development of Miniature VUV Spark Discharge Lamps

a) In general

Generating radiation in the vacuum ultra violet wavelength regime requires the excitation of neutral or ionized atoms/molecules, typically rare gas atoms, into energetically higher atomic/molecular states. Numerous strategies have been applied, such as multi-photon excitation with laser light [111], collisions with accelerated electrons/ions/particles [112-115] or the various electrical discharges [116-120], as used in this work. The generation of VUV light through electrical discharges between two electrodes is a well known process [29]. Spark discharges have been used as early as 1860 by Kirchhoff and Bunsen as an analytical tool for qualitative and later also for quantitative spectral elemental analysis, also known as spark optical emission spectroscopy (OES) [121]. Major advantages of this type of VUV generation are the operational simplicity and the possibility to design lamps in virtually any conceivable geometry. In the following section, three realized lamp designs as well as the development of adequate power supplies will be introduced. A characterization of the discharge and the optical emission properties will be given. Theoretical considerations concerning the discharge mechanisms will be based on works by Paschen [122], Loeb [123], and Druyvesteyn and Penning [124].

b) High voltage-power supplies

The energy source of electrical discharges is an adequate power supply. First attempts were made with high voltage DC/DC converters (E 121, EMCO High Voltage Corporation, Sutter Creek, CA, USA), neon power supplies (VT 12003-120, Ventex Technology Inc, Jupiter, Florida, USA) and flyback transformers, which were recycled from computer monitors. The results from these experiments supported the further development of spark discharge lamps for application in API mass spectrometry [125]. However, they incorporated serious disadvantages such as the uncontrolled discharge frequency in the lower kHz range. The discharge repetition rate was merely determined by self-oscillation of the transformation

4 Results and Discussion

process up to 20 kHz. As a result several lamps suffered from changes of the typical spark characteristics into an arc discharge (see below). This eventually led to strongly elevated electrode temperatures, which the lamp material could not withstand. It turned out that these power supplies were not designed for sustaining controlled, high frequency low impedance discharge pulses. In particular the electronics of the DC/DC converters were severely damaged due to thermal stress after several hours of operation. The first working power supply was realized with a switch-mode power supply (HPE CC400, Hartlauer Präzisions Elektronik GmbH, Grassau, Germany) that was recycled from an exciplex laser. Its actual purpose is to charge capacitors that generate electrical discharges for exciplex production (cf. *3.1.1 Laser Systems*). This high voltage device is capable of providing 14 kV, with 60 mA average current, which is adjustable via pulse width control of the trigger signal. Repeated lamp operation has shown an optimized trigger signal width of 0.18 ms. The repetition rate is adjustable in the range of 1-2000 Hz. In combination with a current back coupling protection module (also recycled from the exciplex laser), an opto-coupler and a digital delay generator a prototype of power supply was constructed. A further modification was necessary since with some lamps the delivered current exceeded the limit of the sparking regime and again the transition into an arc discharge was observed. Therefore a 60 W light bulb was connected in series with the entrance voltage feed as a current limiting device. Upon DC lamp current increase caused by the arc formation, the light bulb briefly flashes. This results in the limitation of the current for the switch-mode supply and subsequently regenerates the typical spark characteristics. In this way a power supply was constructed that provided sufficient voltage, appreciable, variable currents and adjustable repetition rates for sustained spark generation. In combination with the arcing protection system this setup met the demands for the development of spark discharge lamps [125].

Caution: This type of power supply can cause severe damage to the human health in terms of severe electrical shocks!

The final version of the capillary implemented discharge lamp (cf. *3.1.4 Novel APPI Setup*) called for the following requirements for an adequate power supply: (i) 1500 Hz repetition rate, (ii) ~1000 V breakdown voltage and (iii) ~15 mA average output current. Based hereupon a custom high voltage circuit board (DD2010 C-Lader) was designed in cooperation with Hartlauer Präzisions Elektronik GmbH (cf. Figure 30, left). This device allows for adjustable voltages from 0 V - 1500 V DC, adjustable output currents from 0 mA -

4 Results and Discussion

15 mA and a repetition rate of up to 1500 Hz, triggered by an electrical pulse (5 V - 15 V) from an external device.

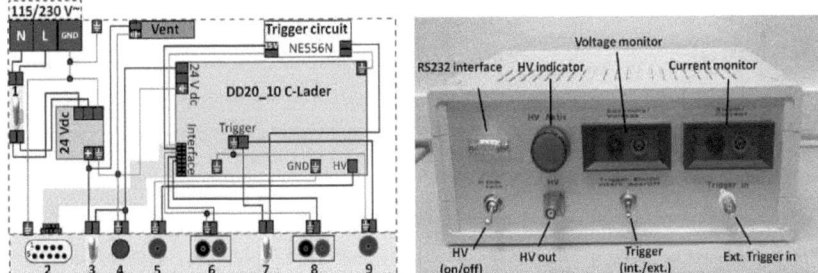

Figure 30: (left) Circuit diagram of the DD20_10 C-Lader power supply with (1) main switch, (2) RS232 interface, (3) HV-on/off switch, (4) HV active indicator, (5) HV-out connector, (6) connectors for voltage monitoring, (7) external/internal-trigger switch, (8) connectors for current monitoring, (9) external trigger-in connector. (right) Photograph of the power supply.

The working principle is based on a switch-mode supply that charges two capacitors (1 pF each) at the HV exit. In short-circuited operation mode the charging process is limited by the breakdown voltage, which is determined by the characteristic lamp setup and its operating conditions (see below). The board requires a 24 V DC supply voltage feed. It is equipped with an interface via a 14 pinout connector supporting several features such as output signals for current or voltage regulation mode, voltage outputs (0-10 V) that are proportional to the average of the discharge current and voltage, voltage inputs (0-10 V) for adjusting the maximal output voltage and current, and an input signal for starting and stopping high voltage generation. Furthermore three pins are assigned to ground and two pins provide a +10 V reference and a +15 V voltage for external use, respectively. The HV device was built into a housing together with a 24 V DC supply, a cooling fan, and a home-built trigger board. The latter uses a NE556N timer, which generates a frequency and the pulse width adjustable trigger signal in the range of 1000 Hz - 2000 Hz and 0.1 ms - 0.2 ms, respectively. Optimum lamp performance was achieved at a trigger pulse duration of 0.18 ms. The voltage feed of the trigger circuit is provided by the +15 V supply voltage pin of the interface (cf. Figure 30, left). The output is connected to the external/internal-trigger switch (cf. Figure 30, left, 7), thus, if desired, an external trigger can be used via the external trigger-in plug (cf. Figure 30, left, 9) as well. On the other hand, the internal pulse generator may also be used for synchronization

4 Results and Discussion

of external devices, by connecting those to the external trigger-in line, but operating the internal trigger. This approach renders synchronization of external devices, such as time-of-flight mass spectrometers, rather simple. The voltage feed of the DD20_10 C-Lader board is provided by a 24 V DC power supply (DPP 50-25, TDK-Lambda Germany GmbH, Aachen, Germany) and can be disconnected via the HV-on/off switch (cf. Figure 30, left, 3). The DC supply itself is running on the 115/230 V AC mains line (cf. Figure 30, left, 1). As soon as the mains switch is turned on the fan is activated. Furthermore, the interface pins for voltage and current monitoring of the high voltage output are connected to the plugs 6 and 8 (cf. Figure 30, left), which, e.g. via a multimeter, render quick control of the discharge characteristics rather handy. The remaining interface pins were connected to the 9-pin sub-D plug (RS232) mounted on the front plate as follows (numbering cf. Figure 30, left, 2): (1) Ground, (2) gives a +15 V pulse if unit is in current regulation mode, (3) capacitor current adjustment by applying a voltage between 0 and 10 V, (4) not connected, (5) gives a +15 V pulse if unit is in voltage regulation mode, (6) output of 10 V reference, (7) capacitor voltage adjustment by applying a voltage between 0 and 10 V, (8) input of a 10 V – 15 V pulse stops and starts generating HV, respectively, and (9) not connected. It is noted that prior to using pins (3) and (7) the corresponding jumpers on the circuit board have to be removed [126].

Caution: This supply can cause health damage in terms of electrical shocks!

In conclusion a versatile, compact, cost efficient (~300 € for the DD20_10 C-Lader) and fairly secure power supply that is perfectly tailored to the developed miniature VUV DC spark discharge lamps has been designed.

c) APPI with or without window

One of the first things to consider when designing lamps that generate radiation in the vacuum ultra violet is the question about the use of windows. In general the implementation of a VUV light source in a mass spectrometric setup is impeded by the necessity of separating the discharge region from the sample gas flow into the mass spectrometer. This is mainly done to prevent the discharge being perturbed or quenched, and when running low pressure discharges to sustain a stable VUV emission. In most cases MgF_2 or LiF windows are used for physical separation. However, these windows restrict the transmission of VUV light below the optical cutoff (MgF_2: 110 nm; LiF: 105 nm) and severely affect the transmission

4 Results and Discussion

efficiencies [127-129]. No window material with lower cutoff wavelength is available. Figure 31 (left) depicts transmission curves for a 5 mm and a 1 mm thick LiF window, derived from data by Korth Kristalle GmbH [130] and by Knop et al. [127], respectively. The extrapolated part of the curve for the 5 mm window clearly shows the steep decrease in transmission from 120 nm down to complete opaqueness at 105 nm. Region 2 marks the normal operating condition for the commercially available Kr-RF low pressure APPI lamp (116.5 and 123.6 nm). Region 1 represents hitherto measured and further expected emissions of the argon operated spark lamp, in which the main part of radiation is located below 130 nm, with a not yet quantified portion below the 105 nm cutoff (OES will be discussed in detail further below). It is clearly seen that a significant portion of the generated radiation cannot be used for ionization, since it is quantitatively absorbed by the window material; in the case of the common APPI lamp (~1 mm window thickness) up to 40 %. Figure 31 (right) shows a direct comparison of operating a spark discharge lamp (design 1; cf. Figure 32, left) on the APPI unit of the LFIS (cf. Figure 24) without and with a window (3 mm thick; with cut flute). Upon irradiating a gas flow of synthetic air with 0.9 ppmV acetone present, as sampled from the photoreactor, a factor of 400 more intense ion signal was obtained with a windowless lamp.

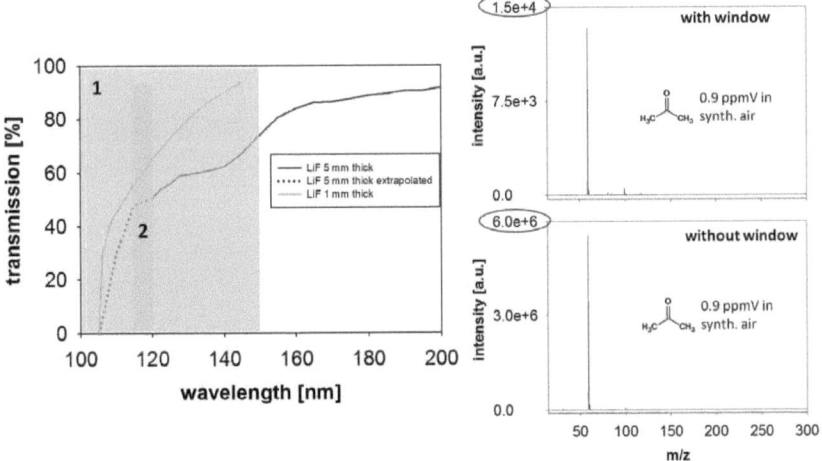

Figure 31: (left) Transmission curves of a 5 mm and a 1 mm thick LiF window. Data were derived from Korth Kristalle GmbH [130] and from Knop et al. [127], respectively. (right) Comparison of mass spectra recorded upon ionization with and without LiF window separation of the spark region of lamp design 1 placed on the APPI unit of the LFIS. As analyte 0.9 ppmV acetone in synthetic air at 1000 mbar were used. The sample gas flow was delivered from the photoreactor.

4 Results and Discussion

Note that the appearance of the two spectra is identical, only the signal intensity differs by orders of magnitude. This is an exceptional result. The question arises as to what caused the poor performance of the window equipped lamp; the strong attenuation of the VUV by the LiF material itself or even further transmission losses by the cut flute. Furthermore, the extent of penning ionization via discharge generated Ar* metastables is not known for this windowless operated lamp. At this point, however, further mechanistic studies were not considered, rather this result was interpreted from an analytical point of view. It was concluded that the further development of miniature spark lamps had to be accomplished without any window material present.

In addition, at medium to high pressure discharge conditions (200 mbar – 1000 mbar), even with the use of high purity argon (99.999%), the surface of the window is generally severely affected due to the presence of reactive species such as H and OH radicals, generated from impurities. This further decreases the VUV transmission efficiency of the window material. Knop et al. have found that the cutoff of LiF shifts to higher wavelengths, up to 107 nm with increasing temperature (134 °C). In this context the heating of the gas within the spark chamber is significant. At common operating conditions for the lamp design 3 (Figure 32, right; discharge chamber pressure $p_{discharge}$ = 719 mbar, 1500 Hz repetition rate) a temperature increase of the glass cover up to 65 °C has been measured. As will be shown in the discussion section about the OES, exactly this emission region (104 nm - 108 nm) is of significant importance of the performance of an Argon discharge lamp.

Consequently, the conclusion was drawn to further develop *windowless* miniature spark discharge lamps.

d) Balanced pressure separation

A novel concept for a quasi physical separation of the discharge gas flow region from the gas flow through the capillary was developed. The pressure $p_{discharge}$ prevailing in the discharge lamp chamber is adjusted closely to match the measured static pressure p_{static} (cf. chapter *4.2.3.1 Characterization of Transfer Capillaries*) of the gas stream through the capillary. As can be seen in Figure 9 an argon flow between 0.1 and 0.5 L·min^{-1} is supplied through the anode. The pumping speed through the cathode is then adjusted up to the point where the condition

4 Results and Discussion

$$p_{discharge} = p_{static} \qquad \text{(eq 25)}$$

is met. This setting creates a virtually separated discharge region without any attenuation of the VUV radiation. UV/VIS calibrated monitoring of the O_2 and N_2 concentrations within the lamp (cf. section: *OES*, further below) offered valuable continuous information about the gas composition *within the discharge region* and it is demonstrated that the windowless approach keeps impurities such as O_2 below 0.2 % even with synthetic air flowing through the transfer capillary. Hence, the mixing of the sample and the discharge gas flow is diffusion limited. To some extend convection, driven by the sparking process, as described, may promote additional mixing, however, only to the maximum extent stated. Along this line the impact of the operating lamp on the main gas stream into the MS was investigated and revealed that the total flow increased by 4 % at most with the discharge running (with $p_{discharge} = p_{static}$). In addition to the generation of VUV light, with the windowless approach deliberate coupling of the gas discharge effluent into the sample gas stream allows further ionization mechanisms to be evoked such as penning ionization. This is easily achieved by raising the discharge chamber pressure above the local sample gas flow pressure:

$$p_{discharge} > p_{static} \qquad \text{(eq 26)}.$$

This highly interesting approach has not been investigated to its full extend yet.

e) Lamp design 1

This preliminary design was used as a radiation source for the APPI unit of the LFIS, later referred to as the "5 ms position". As shown in Figure 32 (left) a high voltage connector was modified with a 0.8 mm thick copper wire, which serves as the ground electrode, and two small Teflon tubes serving as gas in- and outlet lines. Sealing was accomplished by implementing the entire lamp corpus with two suitable rubber O-rings into the aperture of the VUV inlet unit (cf. Figure 24). The lamp discharge channel was positioned parallel to the inner tube of the laminar flow source to ensure maximum illumination. The fairly large electrode distance of 3 mm and the undirected argon gas flow led to a breakdown voltage, which could not be reached by the DD20_10 C-Lader power supply. Thus, the lamp was operated with the HPE CC400 supply only. The discharge stability, long life time of the materials, on/off-reproducibility and the high ionization efficiency when used without the LFIS window, created a valuable reference radiation source.

4 Results and Discussion

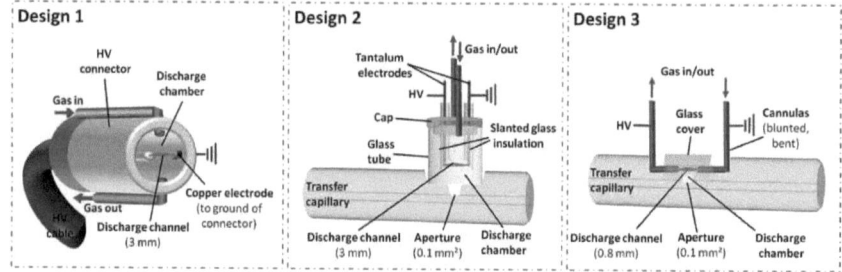

Figure 32: Schematics of home-built spark discharge lamps, (left) design 1, (center) design 2, (right) design 3.

f) Lamp design 2

This more sophisticated design was the first dedicated approach of implementing a windowless spark lamp onto the capillary. As can be seen in Figure 32 (center) it features a separate glass corpus ($\emptyset_{inner}$ = 3.9 mm, $\emptyset_{outer}$ = 5.0 mm and 6 mm height), which is glued onto the one-third grinded capillary, with an optical aperture of 0.1 mm² to the sample flow region. The two-stage cap was made of PVC, with the first stage matching the outer diameter and the second stage matching the inner diameter of the housing. It was sealed with epoxy adhesive. Four holes were conically drilled into the cap, receiving two Teflon tubes ($\emptyset_{inner}$ = 0.8 mm and $\emptyset_{outer}$ = 1.0 mm) for the gas supply and two tips made from glass pipettes ($\emptyset_{inner}$: tapered to ~0.9 mm and $\emptyset_{outer}$: tapered to 1.0 mm) for electrode insulation. Prior to threading the Teflon tubes they were beveled so that the tip diameter matched the hole diameter. For a rather gas tight sealed connection the tubes were subsequently pulled through and the beveled ends were cut even with the inner stage of the cap. The tips of the glass pipettes were slanted and inserted 4 mm deep into the lamp housing with the longer side facing the inner surface of the glass tube. The latter was of essential importance, since earlier attempts without the slanted insulation resulted in spatially instable sparking, in the extreme case with the discharge running along the inner surface of the glass tube. The other end of the insulation protruded around 4 mm out of the cap. The electrodes were made from tantalum rods that were grinded and slightly sharpened ($\emptyset$ = 0.6 mm). The tips of the electrodes were positioned 1 mm before the end of the longer side of the glass insulation. The tip to tip distance, i.e. the discharge channel, was 3 mm, and the distance between the spark region and the sample gas flow of the transfer capillary was 6 mm. Flutes were cut on the other end of the electrodes to wrap around electrical wiring that also end within the glass insulation. Subsequently, heat

4 Results and Discussion

shrinking tubes were imposed on the overlap of the glass insulations with the electrical wires for gas tight sealing, serving additionally as safety measure. For the final step an additional glass tube with the same diameters as the lamp housing was placed on top of the cap. It was filled with epoxy cement and additionally sealed with a heat shrinking tube imposed on both glass cylinders with the cap centered. The lamp chamber could be pumped below 1 mbar. The lamp is safe in terms of electrical shock, it showed acceptable stable spatial and temporal sparking and it featured high irradiation efficiency. This design was mainly used for the "17.5 cm lamp position" (cf. Table 2), also referred to as the "20 µs position", which is spatially as close as possible to the exit port of the transfer capillary, i.e, within the first differential pumping stage (cf. Figure 4, right). To realize ionization at this position, the entire original vacuum interface of the mass spectrometer had to be reconstructed in order to provide the required feed throughs for lamp operation.

Four significant drawbacks of this design are pointed out: (i) Analogue to design 1, the rather large electrode distance, the undirected argon gas flow, and the fairly large volume of the discharge chamber make the operation amenable only to the HPE CC400 supply, (ii) after several hours of operation severe damage of the anode, due to electron bombardment of the surface was observed, (iii) significant electromagnetic perturbation of surrounding devices occurred without sufficient shielding, and (iv) severe wear of the used materials became apparent, due to exposure to VUV light, chemically reactive species and thermal stress. Eventually gas leakages were noticed and in one of the worst cases cracking of the glass insulations occurred.

g) Lamp design 3

The final design is the technically most mature and its development considered most of the mentioned drawbacks of design 2. A detailed description has already been given in section *3.1.4 Novel APPI Setup*. The essential improvements are (i) the smaller electrode distance (0.8 mm vs. 3 mm), (ii) the directed argon flow from the anode to the cathode, (iii) the significant smaller discharge volume (8.3 mm^3 vs. 50 mm^3), (iv) the smoother and more evenly shaped electrode surface, and (v) the significantly reduced distance between the spark gap and the sample flow region of the transfer capillary (0.8 mm vs. 6 mm). As result, (i) the breakdown voltage is lowered from > 1500 V to < 1000 V, (ii) a cooling effect of the anode occurs, (iii) a drastic reduction of anode ablation is observed, (iv) the discharge region

4 Results and Discussion

is significantly more confined, (v) the sample gas stream is more precisely and efficiently illuminated, (vi) the spark-to-spark stability is increased and, (vii) operation with the smaller, more safe (in terms of hazardous electrical shocks) and cost efficient DD20_10 C-Lader power supply is possible. With this setup electromagnetic perturbation of surrounding devices was not observed. Despite the reduced size discharge region, at least identical ionization efficiencies were observed, most probably resulting from the precise spatial alignment of the radiating spark region with the sample gas flow. The construction of this lamp type was rather straight forward. The flutes and holes were machined on a computer controlled milling machine; the cannulas were blunted, bent and on one end imposed with Teflon tubes exactly matching the outer diameter of the cannulas. All components (capillary, cannulas, electrical wiring, and the glass cover) were placed into a Teflon mold and filled up with cement resulting in a compact, sealed and safe design. Care had to be taken when positioning the tips of the electrodes, such that the protrusion into the discharge chamber was sufficient (~0.5 mm) but without contacting the chamber walls. Otherwise spatial spark instabilities arose, analogue to lamp design 2. With design 3 no wear of the used materials has been hitherto observed, except a slight color change of the anode.

h) Operating stability tests

A long-term stability test over ten hours was performed with lamp design 3 on position "9 cm" of the transfer capillary (cf. Table 2). Typical operating conditions were applied: The argon flow rate was set to 0.5 $L \cdot min^{-1}$, the repetition rate was 1500 Hz, and the discharge pressure was closely matched to the local static pressure. The DD20_10 C-Lader was used as the power supply. The sample gas was provided by the photoreactor with acetone as analyte (0.3 ppmV in synthetic air at 1000 mbar). The pressure in the reactor was balanced with a continuous flow of synthetic air matching the amount sampled. The resulting signal fall-off within the recorded ion chromatogram closely followed the expected signal loss solely due to dilution. Consequently, the lamp did not show any noticeable discharge changes that might have affected the ionization efficiency over these ten hours of continuous operation. Furthermore, several tests concerning the lamp on/off stability were performed and again, very good performance was observed. In conclusion, this lamp design nicely meets the requirements of a VUV radiation source applied in API MS, particularly with respect to radiation stability and reproducibility.

4 Results and Discussion

i) Determination of lower detection limits (LODs)

For the LOD experiments the ion trap was run in the same mode as for typical degradation product studies. The MS was operated in the alternating mode, with 5 ms and 20 ms accumulation time for the positive and negative mode, respectively. One data point in the chromatogram was the result of 10 averaged single spectra, thus every 250 ms one data point for each mode was recorded. The mass range was set from m/z 15 to m/z 500. For ionization lamp design 3 was used on position "9 cm" of the transfer capillary and operated as described above. The experimental procedure for the LOD investigations was as follows: The lower detection limit for benzene was determined by stepwise injecting a 0.1 M solution in acetonitrile into the photoreactor. The LOD for 2-butanone was established by combining a carrier gas flow of synthetic air with a minor flow of the oxygenated gas mixture containing 2-butanone at a mixing ratio of 96 ppbV in nitrogen. The main gas supply was accomplished through a mass flow controller (1179A Mass-Flo-Controller, 2000 sccm; MKS Instruments, Andover, MA, USA), which was connected to the sheath gas inlet of the transfer unit (cf. Figure 12, 4). The minor flow was connected to the entrance of the transfer unit (cf. Figure 12, 5) via a 10 sccm mass flow controller (1179A Mass-Flo-Controller, 10 sccm; MKS Instruments, Andover, MA, USA). Gas flow calibrations were performed with either a wet meter TG05 (Ritter Apparatebau GmbH and Co. KG, Bochum, Germany) or a home-built bubble counter, respectively. Both mass flow controllers were connected to a 647 C multi-gas-controller unit (MKS Instruments).

The LODs for benzene ($[M]^+$, m/z 78), and 2-butanone ($[M+H]^+$, m/z 73) were estimated following the procedure described by Kaiser and Specker [131]:

$$x_{LOD} = \frac{3\sigma_s}{b_s} \qquad (\text{eq } 27)$$

with x_{LOD} as the lowest analyte concentration being detected with a statistical confidence of 99.7 %, σ_s as the standard deviation of several measurements without analyte present and with b_s the slope of the calibration line. The standard deviations were determined in a five minute run of a single ion chromatogram with no analyte present. The slopes were calculated from the calibration curves shown in Figure 33. LODs of 0.5 ppbV for benzene and 0.1 ppbV for 2-butanone were obtained. Hence, the obtained lower detection limits were sufficient for the intended degradation studies, since degradation product mixing ratios in the upper ppbV range were expected.

4 Results and Discussion

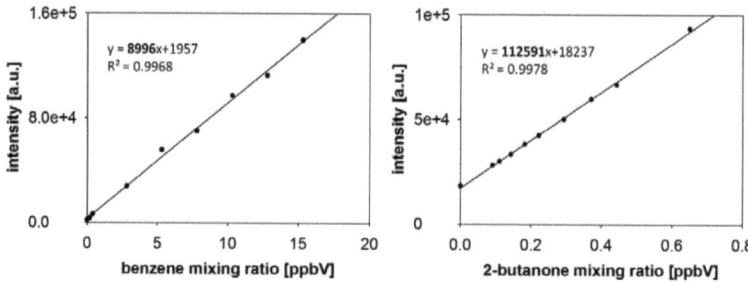

Figure 33: Calibration curves for the determination of the lower limit of detection (LOD) with the novel APPI setup. The analytes are (left) benzene and (right) 2-butanone, respectively.

It is pointed out that the optimization of the design and the operating conditions of the discharge lamp has not been accomplished to its full extend yet and thus even better performance is expected in the near future.

j) Experimental sparking characteristics of design 3 with the DD20_10 C-Lader

The appearance of a single spark in lamp design 3 is characterized by the formation of several stream lines, which circular surround the electrode edges. This effect leads to the unexpected high brightness of the discharge. Hence, for an optimum ionizing photon flux yield the electrode tips should not be of very sharp shape to obtain high electrical field gradients as required for example in corona discharges. The present operating conditions are providing sufficient field strength, cover the entire aperture to the bulk gas flow of the capillary, and conserve the spatial and temporal spark-to-spark stability. Along this line a slight drawback of the circular discharge shape has been recognized. Since photons generated on the upper edges of the cannulas apparently have to travel a longer path to the sample gas flow, the probability entering the aperture is decreased; both due to the radial distribution and to absorption processes. A new concept for the electrode shapes offered first promising results in terms of directing the photon flux, and higher temporal and spatial spark-to-spark stability. Here, cannula tips were one-third grinded in 0.5 mm length and the sharp edges were evenly flattened. The electrodes were arranged exactly facing each other, with the argon gas stream flowing over the tips. A flat surface of several stream lines resulted between the electrodes.

4 Results and Discussion

This bright surface may carefully be aligned on top of the aperture to the capillary. However, this design has not been tested within a transfer capillary yet.

Typical operating conditions were applied for the following spark investigations: The radiation source was implemented on "position 9 cm" of the transfer capillary (cf. Table 2), with $p_{static} = p_{discharge}$ =719 mbar. The argon flow was set to 0.5 L·min^{-1}, the repetition rate was 1500 Hz and the trigger signal duration was adjusted to 0.18 ms. Ambient air at 997 mbar upstream pressure was flowing through the transfer capillary.

Figure 34 (a) depicts several discharge cycles with the quantified temporal current and voltage evolution (for the experimental setup see also *3.1.4.2 Characterization of the Discharge Lamp*). The experimentally determined breakdown potential is V_{break} = 860 V and the maximum pulse current is up to 2.2 A. During the spark event the cathode voltage rapidly drops to a minimum of 60 V at which the discharge eventually becomes unstable and stops.

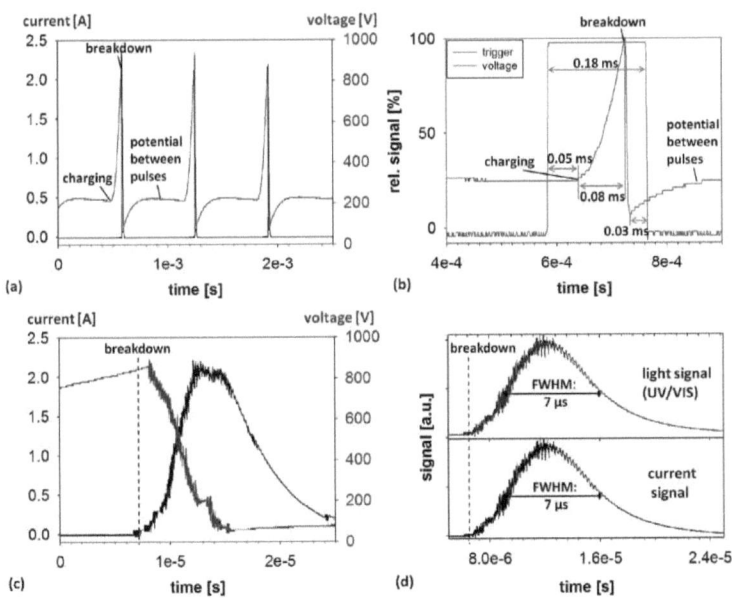

Figure 34: Spark discharge characteristics of lamp design 3 operated with the DD20_10 C-Lader. (a) Temporal correlation of the current and voltage evolution, illustrated by three subsequent breakdowns, (b) temporal correlation of the trigger signal and the potential on the cathode, (c) temporal evolution of the voltage and current during one spark, and (d) temporal correlation of the current and the light emission in the UV/VIS during one spark.

4 Results and Discussion

Figure 34 (b) illustrates the time dependent correlation between the driving voltage of the cathode and the trigger signal during one spark event. As can be seen 50 µs after the positive slope of the trigger signal the electronic circuit of the DD20_10 C-Lader supply starts charging the capacitors with approximately $1.3 \cdot 10^7$ V·s^{-1}. After 80 µs the voltage has reached the breakdown potential[2] and the two capacitors discharge within 8 µs. Note, that the trigger signal is still high during the sparking process, which means that the switch-mode supply continuously recharges the capacitors. This eventually leads to additionally produced current than just the load of the two capacitors would generate. The discharging process, however, is faster than recharging, so that the overall voltage drops within 8 µs below the discharge threshold level. The trigger signal is still high for further 30 µs which recharges the capacitors to about 25 % of the breakdown voltage (~220 V). The temporal variation of the breakdown cycles, relative to the rising slope of the trigger signal, was measured to be within ± 0.5 µs accuracy, which makes this lamp a valuable tool for applications where a precisely triggered, pulsed radiation source is required.

A close up of the voltage and current curves is shown in Figure 34 (c). The well defined starting point of the spark event is clearly discernible. The recorded current curve is of Gaussian shape with its maximum close to the minimum of the cathode voltage and with a FWHM of around 7 µs, as marked in Figure 34 (d). The latter panel also shows the temporal evolution of the light signal obtained for the UV/VIS region with fairly good overlap of both curves. Thus the FWHM duration of the light pulse is 7 µs.

k) Theoretical considerations on the spark characteristics

As simple as sparking devices are, as complex may theoretical considerations about the prevailing mechanisms become. Thus the purpose of this section is a rough description based on some well known theoretical approaches described in the literature. The first fundamental work on spark mechanisms was presented by Paschen in 1889 who investigated the breakdown voltage V_{break} in dependency on the product of electrode distance d_{elec} and the discharge pressure $p_{discharge}$ [122]. A typical Paschen curve for argon in a double logarithmic plot of V_{break} and the product of $p_{discharge}$ and the electrode distance d_{elec} is shown in Figure 35, derived from data points given by Druyvesteyn et al. [124] for plane, parallel iron plates as

---

[2] "[...] the voltage for which the initial current is increased by a very large factor [...] is called the *breakdown potential* [...]." [124]

4 Results and Discussion

electrodes within a static gas. The present operating conditions are marked in the plot suggesting a breakdown voltage of 900 V, which is in very good accordance to the experimentally measured V_{break} = 860 V. For simplification purposes, the present setup will thus be theoretically treated as an assembly of two parallel electrodes, placed within a static gas.

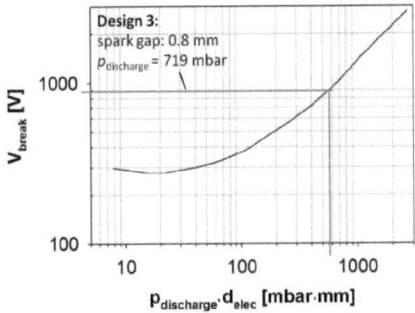

Figure 35: Typical Paschen curve for argon, derived from data given by Druyvesteyn and Penning [124]. The sketched lines illustrate the operating conditions of lamp design 3.

First, the situation prior to the first single spark event is considered. The resistance of the gap is high, preventing electrons from leaving the cathode to the anode even with a moderately voltage applied. However, as stated by Loeb [123]:"If no electrons are produced there is no current". However, there are always free electrons present in the gas due to cosmic rays and natural radioactivity. Any potential gradient between the electrodes immediately induces a small base current i_0. Due to acceleration within the electrical field the electrons are capable of generating secondary products (ions, excited atoms and photons) on their path to the anode. The probability of the primary electrons to cause ionization along their travel path within a specific gas species is expressed by the first Townsend coefficient α [m^{-1}], which is calculated as [132]:

$$\alpha = \frac{1}{\lambda_e} \cdot e^{\frac{E_i}{E_e}} \quad \text{(eq 28)}$$

with λ_e = electron mean free path [m], E_i = ionization energy of the gas [eV], and E_e = average collision energy [eV]. Equation 28 expresses the product of the number of collisions per unit length and the ionization probability per collision. E_e is calculated according to [132]:

4 Results and Discussion

$$E_e = \lambda_e \cdot E \qquad (eq\ 29)$$

with E = electric field [V·m^{-1}], which is calculated assuming a linear dependence. The electron mean free path is derived from [132]

$$\lambda_e = \frac{k_b \cdot T}{\sigma_{ei} \cdot \mathcal{P}_{discharge} \cdot 100} \qquad (eq\ 30)$$

with k_b = Boltzmann constant [J·K^{-1}], T = temperature [K], σ_{ei} = electron ionization cross section [m²], and $p_{discharge}$ [mbar]. Once those secondary products are formed they are capable of liberating electrons from the cathode, due to positive ion impact, photoelectron emission, and partially due to the impact of metastables [124]. These secondary electrons also travel within the electrical field E from the cathode to the anode and generate further ions, photons, and metastables resulting in an avalanche establishing the self sustained base current i_0 and eventually lead to the observed breakdown. As can be seen in Figure 34 (c), the capacitor charge (and the additional charge still delivered from the power supply) is injected from the cathode into the plasma, resulting in a rapidly increasing current and decreasing voltage. The latter process induces an electrical field change within a spark event, accounting approximately to

$$\frac{dE}{dt} \approx 1.9 \cdot 10^{10} \left[\frac{V}{m \cdot s}\right] \qquad (eq\ 31).$$

At around $E = 7.5 \cdot 10^{-4}$ [V·m^{-1}] (calculated with the lowest measured voltage[3]), newly liberated electrons from the cathode experience a field strength that does not lead to a collision energy E_e for sufficient formation of new ions, excited atoms, and/or photons. The discharge becomes unsustainable and stops.

At this point the impact of the repetitive spark operation mode (every 0.7 ms one breakdown event) and the continuous gas exchange of the discharge chamber is qualitatively implemented into the presented model. The gas flow (0.5 L·min^{-1}) from the anode to the cathode leads to approximately 50 % gas exchange in the discharge chamber (8.3 mm³) in between two spark events. Hence, the prevailing operating conditions provide elevated stable primary electron and positive ion background concentrations than naturally occurring. The gas flow direction, as mentioned, has significant impact on the single temporal and spatial spark stability. This is readily explained by the presented model, since the liberation of a new

---

[3] Not accounted are space charge effects that might change the effective field strength experienced by an electron within the gap.

4 Results and Discussion

generation of electrons is enhanced by the fluid dynamical transport of ions and excited atoms towards the cathode. The directed flow also improves the spark-to-spark stability, as the impact of positive ions possibly liberates electrons from the cathode in an early stage of the discharge event. In accordance with this picture also a lower breakdown voltage is expected contrary to the model of independently occurring single spark events. Furthermore, the probably orders of magnitude higher background concentration of primary electrons and the permanent plateau potential of 220 V between the pulses (cf. Figure 34, a) raise the base current i_0 to a much higher level. This also evokes a decrease of the breakdown voltage and an increase in the temporal stability of the spark ignition. However, without sufficient dilution, i.e., an even higher charge carrier background level, there is risk of creating a conducting gap with nearly no resistance, which would result in extremely high currents and eventually burnout the lamp. In that case all the electrons provided by the switch-mode supply would not be used to charge the capacitors back again but to directly migrate through the gap. This scenario describes the turning point from a spark into an arc discharge. Thus, dilution is a balancing act between preventing the accumulation of too many charge carriers but still providing appreciable high concentrations of primary electrons and positive ions for optimum lamp performance.

Up to now the fate of an electron on its passage to the anode has barely been mentioned. However, this is what keeps the discharge running in terms of the generated secondary products and this is what makes the discharge to a valuable VUV radiation source for APPI-MS applications. As described by Druyvesteyn and Penning [124] there are basically five processes occurring while the electrons travel within an electrical field through a gas of a given pressure $p_{discharge}$: (i) Elastic collisions with gas atoms/molecules, (ii) electronic excitation of atoms/molecules, (iii) vibrational excitation of molecules, (iv) ionization of atoms/molecules, and (v) gain of electron kinetic energy. The sum of all fractional energy contributions for each process is expressed by the electron energy balance equation [124]. For argon as a monatomic gas and assuming that the impact of impurities such as O_2 and N_2 are insignificant, the electron energy loss due to vibrational excitation cannot be present. Furthermore, it is stated [124] that in case of the prevailing discharge conditions the impact of elastic collisions is to a first approximation negligible. These assumptions reduce the balance equation to

$$1 = \eta \cdot E_i + \eta \cdot E_e + \sum_n \xi_{el.exc;n} \cdot E_{el.exc;n} \quad \text{(eq 32)}$$

4 Results and Discussion

with $\eta \cdot E_i$ and $\eta \cdot E_e$ describing the fractions of the electron energy used for ionization and for the increase of the electron kinetic energy, respectively. The last term on the right side of equation 32 expresses the fraction of the total available energy that is lost in electronic excitation. E_i represents the ionization energy of argon in [eV] and E_e is calculated according to equations 28 and 29. η is the ionization coefficient [V^{-1}] and describes the average number of ionization processes per volt potential difference passed by the electron [124] and is related to the Townsend coefficient in the following way

$$\eta = \frac{\alpha}{E} \qquad \text{(eq 33).}$$

The coefficient $\xi_{el.\ exc;\ n}$ [V^{-1}] expresses the probability of exciting the nth electronic level and $E_{el.\ exc;\ n}$ [eV] represents the energy of the nth electronically excited level. According to equation 32 the fractional contribution of each process can now be calculated and is typically plotted versus $E \cdot p_{discharge}^{-1}$. Experimental values of η versus the product of electrical field and discharge pressure in argon were taken from Druyvesteyn and Penning [124]. The ionization energy of argon is E_i = 15.7 [eV]; the average electron energy E_e was calculated according to equations 29 and 30, with an ionization cross section of σ_{ei} = 3.5·10^{-20} m^{-2} [133]. The contribution of electronic excitation is derived from equation 33 since the other two terms are determined. The resulting half-logarithmic plot is shown in Figure 36; the region of the discharge conditions using lamp design 3 is shaded red. As can be seen the contribution of electron kinetic energy increase is negligible throughout the entire breakdown.

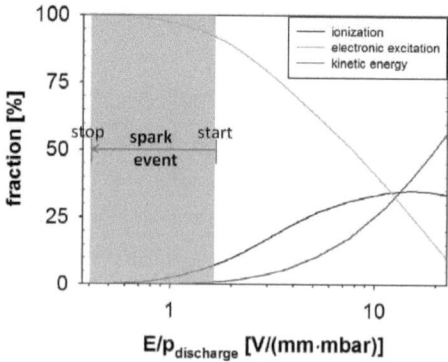

Figure 36: Mean partial electron energy contributions to the processes of ionization of the discharge gas, electronic excitation of the discharge gas, and kinetic energy increase of the electron stream within a discharge in pure argon. Highlighted is the operating region of the prevailing discharge characteristics in lamp design 3.

4 Results and Discussion

At the beginning of a spark event the mean fractional loss due to ionization accounts for around 5 %, the rest, nearly 95 %, is driving electronic excitation. The $E \cdot p_{\text{dischrage}}^{-1}$ values decrease from 1.5 to 0.1 $V \cdot mm^{-1} \cdot mbar^{-1}$ along with the rapid voltage drop from 860 V to 60 V (cf. Figure 34 c). Also the fractional loss due to ionization drops to zero at the end of the spark and the fractional impact of electronic excitation climbs to 100 %. At this point it is again noteworthy that this picture can only present an estimate of the relative contributions between electron kinetic energy increase, ionization, and electronic excitation within the electron energy balance of a discharge in argon. It does not account for losses due to elastic collisions or the accompanied Ramsauer effect [134] at low electron energies[4]. Furthermore, impurities such as O_2 and N_2 would change the situation by means of the fractional loss due to possible vibrational excitation, which then needs to be taken into account. Additional impurities with lower ionization potentials than the energy of argon metastables (11.55 and 11.72 eV; cf. Table 3) would further change the relative contribution between ionization and electronic excitation, since in subsequent Penning ionization reactions [124] the primary electronic excitation would be converted to charged species. Nevertheless, Figure 36 clearly shows that excitation is by far the dominating process within the discharge operating conditions of lamp design 3. Photons or metastables are essentially responsible for the liberation of new electrons from the cathode; both are the dominating energy carriers to sustain the discharge. This finding supports the directed flow approach, since an effective transport of metastables is thus ensured. The large degree of electronic excitation within the miniature spark discharge chamber is the explanation for the efficient generation of VUV photons. This lamp design is thus a very attractive alternative VUV radiation source for APPI-MS applications.

l) Optical emission spectroscopy (OES)

It was shown that the discharge efficiently generates electronically excited species. Based on optical emission spectroscopic investigations this section discusses the determination of these species and their possible fates within a comprehensive discharge chemistry. Figure 37 shows optical emission spectra of the spark discharge lamp in the range of 105 nm to 1100 nm. Note that wavelength dependent transmission discrimination of the fiber optics and the LiF window, due to constraints of the experimental setup (cf. *3.1.4.2*

---

[4] This effect is based on the wave-particle dualism of the electron. It shows that the collision probability is a function of the electron energy. This probability is characterized by a maximum when the electron energy with its corresponding de-Broglie wavelength comes close to the dimensions of the surrounding gas species.

4 Results and Discussion

Characterization of the Discharge Lamp), are not considered. The UV/VIS spectrum in Figure 37 (right) pictures the emission of the discharge lamp operated as discussed before ($p_{static} = p_{discharge}$ = 719 mbar; 1500 Hz repetition rate, trigger pulse width = 0.18 ms using the DD20_10 C-Lader power supply; ambient air at 997 mbar as the sample gas stream through the transfer capillary). The spectrum is dominated by characteristic Ar I lines (argon in the neutral state), of which the most intense ones are listed in Table 3. The Grotriam diagram for Ar I [135] reveals that many of these lines between 390 and 1100 nm are precursor emissions for transitions in the VUV (Ar*(^{2}[3/2]* J=1 $\rightarrow$ $^1S_0$ = 106.7) and Ar*(^{2}[1/2]* J=1 $\rightarrow$ $^1S_0$ = 104.8 nm)) and for the generation of excited argon species with forbidden transitions to the ground state (Ar*(^{2}[3/2]* J=0) = 11.55 eV and Ar*(^{2}[1/2]* J=0) 11.72 eV). In principal this allows for monitoring the UV/VIS light intensity as a measure of the generated VUV radiation and metastable formation. The spectrum further shows the emissions from electronically excited N_2 (6 bands from 280 - 400 nm), O I (777 nm) and H I (656 nm), which resulted from diffusion of the sample gas flow into the discharge region. The air concentration within the argon atmosphere amounted to 0.8 %, which was determined by calibrating the relative intensity of the O I (777 nm) and Ar I (763 nm) lines with synthetic air. The broad continuum between 250 and 700 nm is tentatively assigned to free-free and free-bound transitions of the electrons.

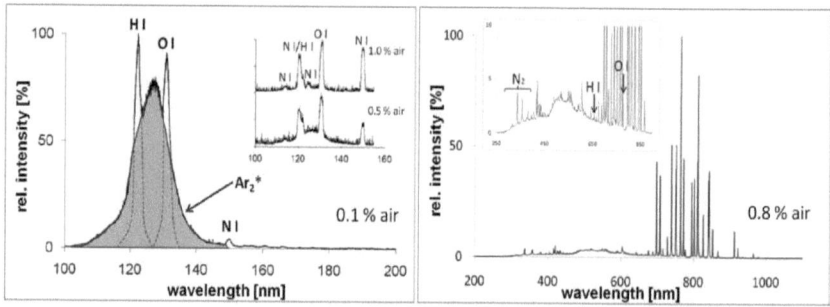

Figure 37: (left) VUV spectra of the spark discharge in 1000 mbar argon with 0.1, 0.5 and 1.0 % synthetic air present and (right) UV/VIS spectrum of the spark discharge in 719 mbar argon with 0.8% synthetic air present.

The VUV spectrum at an argon pressure of 1000 mbar with 0.1 % air present (cf. Figure 37 left) is characterized by the second excited argon dimer Ar_2^*($^1\Sigma_u^+$ and $^3\Sigma_u^+$) continuum

4 Results and Discussion

centered at 127 nm, with a line width of FWHM = 20 nm, and three lines that are assigned to H I (121.5 nm)/N I (119.9, 120.2 and 120.7 nm), O I (130.2, 130.4 and 130.6 nm) and N I (149.2 and 149.5 nm) transitions. The lines at 106.7 and 104.8 nm, as predicted from the UV/VIS precursor emissions, do not appear. The VUV radiation is most likely completely absorbed by the LiF window due to its spectral cutoff (cf. Figure 31). The absence of the $Ar^*(^2[3/2]^*\ J=1)$ emission to the ground state is also partially assigned to the reduced transmission of the window, but moreover this state is primarily reacting to form argon excimers $Ar_2^*(^1\Sigma^+_u)$ with argon ground state $Ar(^1S_o)$ atoms, as described by Millet et al. [114]:

$$Ar^*(^2[3/2]^*\ J=1) + 2\ Ar(^1S_o) \rightarrow Ar_2^*(^1\Sigma^+_u) + Ar(^1S_o) \quad \text{(rxn 12)}$$

The authors further proposed a pathway to form the triplet state of the argon excimer $Ar_2^*(^3\Sigma^+_u)$ within the discharge chemistry:

$$Ar^*(^2[3/2]^*\ J=0) + 2\ Ar(^1S_o) \rightarrow Ar_2^*(^3\Sigma^+_u) + Ar(^1S_o) \quad \text{(rxn 13)}$$

Here the metastable state $Ar^*(^2[3/2]^*\ J=0)$ at 11.55 eV reacts in a three body collision process with argon atoms in the ground states to generate the excited dimer. It is speculated that in the present setup the metastable state itself is not participating in any analyte ionizing process (in terms of ionizing analyte molecules within the transfer capillary), since the discharge region is virtually separated from the sample gas flow (see above) and thus collisions with the analyte are not likely to occur. However, reaction 13 turns this inactive species into a VUV radiating excimer, hence an active analyte ionizing species. Taking into account the relative intense precursor emission (~30 %, cf. Table 3) in the UV/VIS leading to this metastable state, this reaction appears to be very important for the total ionization efficiency of this lamp type. The radiative transition of both excimer states to the dissociative ground state is centered at 127 nm. The reported lifetimes are $\tau_{singlet} = 6$ ns and $\tau_{triplet} = 2.86\ \mu s$ for $Ar_2^*(^1\Sigma^+_u)$ [136] and $Ar_2^*(^3\Sigma^+_u)$ [114], respectively. Due to the long lifetime of the triplet state, resonant energy transfer of the excimer to O I, N I and H I is enhanced, as, e.g., described for oxygen by Moselhy et al. [137]. The spectra in Figure 37 (left) with 0.5 and 1.0 % air in 1000 mbar argon clearly demonstrate the quenching effect with small amounts of impurities. The excimer continuum completely disappears and H I (121.5 nm), O I (130.2, 130.4 and 130.6 nm) and N I (113.4, 119.9, 120.2, 120.7, 124.3, 149.2 and 149.5 nm) lines are enhanced. The integrated spectral intensity from 105 to 155 nm remains constant, though. This result is supported by an investigation of the ionization efficiency regarding acetone in which O_2 was

4 Results and Discussion

added stepwise to the argon atmosphere of the lamp. The MS signal stayed virtually constant even with up to 10 % oxygen present. Pure oxygen, however, caused a considerable decrease in the MS signal, apparently due to a significant change in the discharge characteristics itself.

Table 3: Observed Ar I emission lines between 200 and 1100 nm. The intensity is relative to the sum of the Ar I line intensities. Non precursor transitions are denoted with x. For the precursor emission lines the term of the lower state of that transition and its energy to the Ar ground state ($^1S_0$) are listed, with *(104.8 or 106.7 nm)* and *(m)* indicating a radiative or metastable state, respectively. Not definable intensities are denoted with n.d.

wavelength lit.[nm] [135]	observed wavelength [nm]	relative intensity [%]	term of the lower excited Ar* state of the precursoremission	Δ energy Ar* - Ar($^1S_0$) [eV]	
394.8979	394.9	0.204	$^2[3/2]^*$ J=0	11.55	*(m)*
404.4418	404.4	0.240	$^2[3/2]^*$ J=1	11.62	*(106.7 nm)*
415.859	415.9	0.464	$^2[3/2]^*$ J=0	11.55	*(m)*
420.0674	420.0	0.601	$^2[3/2]^*$ J=0	11.55	*(m)*
425.9362	426.0-427.2	n.d.	$^2[1/2]^*$ J=1	11.83	*(104.8 nm)*
426.6286	426.0-427.2	n.d.	$^2[3/2]^*$ J=1	11.62	*(106.7 nm)*
427.2169	426.0-427.2	n.d.	$^2[3/2]^*$ J=1	11.62	*(106.7 nm)*
430.0101	430.1	0.266	$^2[3/2]^*$ J=1	11.62	*(106.7 nm)*
433.3561	433.4	n.d.	$^2[1/2]^*$ J=1	11.83	*(104.8 nm)*
433.5338	433.4	n.d.	$^2[1/2]^*$ J=1	11.83	*(104.8 nm)*
434.5168	434.7	0.308	$^2[1/2]^*$ J=1	11.83	*(104.8 nm)*
451.0733	451.1	0.224	$^2[1/2]^*$ J=1	11.83	*(104.8 nm)*
452.2323	452.2	0.201	$^2[1/2]^*$ J=0	11.72	*(m)*
459.6097	459.6	0.200	$^2[1/2]^*$ J=1	11.83	*(104.8 nm)*
516.2285	516.2	0.443	x	x	
518.7746	518.7	0.473	x	x	
522.1271	522.1	0.406	x	x	
545.1652	545.1	0.403	x	x	
549.5874	549.7	0.483	x	x	
555.8702	555.9	0.468	x	x	
557.2541	557.3	0.414	x	x	
560.6733	560.7	0.447	x	x	
565.0704	565.0	0.372	x	x	
573.952	574.0	0.305	x	x	
583.4263	583.4	0.265	x	x	
586.031	586.0	0.265	x	x	
588.8584	588.9	0.320	x	x	
591.2085	591.3	0.324	x	x	
592.8813	592.9	0.290	x	x	
594.2669	594.3	0.272	x	x	
598.7302	598.8	0.260	x	x	
603.2127	603.3	0.558	x	x	
605.9372	606.0	0.359	x	x	

4 Results and Discussion

610.5635	610.6	0.285	x		x	
617.3096	617.5	0.254	x		x	
621.2503	621.4	n.d.	x		x	
621.5938	621.4	n.d.	x		x	
629.6872	629.8	0.190	x		x	
630.7657	630.9	0.197	x		x	
636.9575	636.9	0.173	x		x	
638.4717	638.6	0.183	x		x	
641.6307	641.9	0.255	x		x	
653.8112	653.9	0.173	x		x	
660.4853	660.5	0.165	x		x	
666.4051	666.5	0.159	x		x	
667.7282	667.7	0.204	$^2[3/2]^*$ J=1		11.62	(106.7 nm)
675.2834	675.0-675.6	n.d.	x		x	
675.6163	675.0-675.6	n.d.	x		x	
687.1289	687.2	0.310	x		x	
688.8174	688.8	0.166	x		x	
693.7664	693.8	0.193	x		x	
696.5431	696.6	5.111	x		x	
703.0251	703.1	0.275	x		x	
706.7218	706.8	n.d.	$^2[3/2]^*$ J=0		11.55	(m)
706.8736	706.8	n.d.	x		x	
714.7042	714.7	0.472	$^2[3/2]^*$ J=0		11.55	(m)
720.698	720.8	0.138	x		x	
727.2936	727.4	1.080	$^2[3/2]^*$ J=1		11.62	(106.7 nm)
731.6005	731.5	0.129	x		x	
735.3293	735.4	0.178	x		x	
738.398	738.4	6.123	$^2[3/2]^*$ J=1		11.62	(106.7 nm)
750.3869	750.4	4.412	$^2[1/2]^*$ J=1		11.83	(104.8 nm)
751.4652	751.5	6.208	$^2[3/2]^*$ J=1		11.62	(106.7 nm)
763.5106	763.6	11.676	$^2[3/2]^*$ J=0		11.55	(m)
772.3761	772.4	n.d.	$^2[3/2]^*$ J=0		11.55	(m)
772.4207	772.4	n.d.	$^2[1/2]^*$ J=0		11.72	(m)
794.8176	794.9	4.176	$^2[1/2]^*$ J=0		11.72	(m)
800.6157	800.6	2.234	$^2[3/2]^*$ J=1		11.62	(106.7 nm)
801.4786	801.5	4.333	$^2[3/2]^*$ J=0		11.55	(m)
810.3693	810.4	5.053	$^2[3/2]^*$ J=1		11.62	(106.7 nm)
811.5311	811.5	10.017	$^2[3/2]^*$ J=0		11.55	(m)
826.4522	826.4	2.239	$^2[1/2]^*$ J=1		11.83	(104.8 nm)
840.821	840.9	3.990	$^2[1/2]^*$ J=1		11.83	(104.8 nm)
842.4648	842.5	4.623	$^2[3/2]^*$ J=1		11.62	(106.7 nm)
852.1442	852.2	1.558	$^2[1/2]^*$ J=1		11.83	(104.8 nm)
860.5776	860.6	0.024	x		x	
866.7944	866.8	0.374	$^2[1/2]^*$ J=0		11.72	(m)
912.2967	912.3	1.429	$^2[3/2]^*$ J=0		11.55	(m)
919.4638	919.5	0.025	x		x	

4 Results and Discussion

922.4499	922.5	0.481	$^2[1/2]^*$ J=1		11.83	*(104.8 nm)*
929.1531	929.2	0.013	x		x	
935.422	935.4	0.055	$^2[1/2]^*$ J=1		11.83	*(104.8 nm)*
965.7786	965.7	0.149	$^2[3/2]^*$ J=1		11.62	*(106.7 nm)*
978.4503	978.5	0.059	$^2[1/2]^*$ J=1		11.83	*(104.8 nm)*
1047.0054	1047.1	0.005	$^2[1/2]^*$ J=0		11.72	*(m)*
1047.8034	1047.8	0.005	x		x	

m) VUV emission efficiency in comparison to the commercially available APPI lamp

Measurements with the commercially available krypton radio frequency (Kr-RF) lamp showed that the total VUV radiation (> 8.3 eV) of the home-built spark discharge lamp amounts to 75 % of the Kr-RF lamp emission. It is noteworthy that the radiation at 104.8 nm (precursor emission accounts for 13.5 %, cf. Table 3) and discrimination of wavelengths below 116 nm due to the LiF window of the VUV spectrometer are not considered. In the windowless lamp design this radiation is quantitatively available for ionization.

n) VUV emission spectroscopy below 105 nm

As stated above, the significant drawback of the VUV spectroscopic setup was the need of a physical separation of the discharge from the evacuated chamber containing the grating and the detector (MCPs). However, as discussed in depth earlier, the LiF transmission efficiencies and the cutoff merely revealed parts of the actually present VUV emission. In a cooperation with Resonance Ltd. (Barrie, ON, Canada) spectroscopic investigations without the use of window material became possible. A modified diode array detector was used that is sensitive down to 30 nm and that allows for operation at elevated pressures. The spark setup in terms of electrode distance and gas flow was analogue to lamp design 3, however, built into a mount fitting the flange of the spectrometer. In a first attempt helium (830 mbar) was employed as the background gas of the spectrometer and mixtures of helium and argon were used for the gas flow through the electrodes. High voltage was supplied by the DD20_10 C-Lader, operated as described before. Two VUV spectra with 1 % Ar in He and 9.5 % He in Ar present, respectively, are shown in Figure 38: (left) and (right). The left spectrum shows a broad Ar I emission which is speculated to be a combination of the unresolved 104.8 and 106.7 nm emissions. The well known emission lines of N I, O I and H I are also present in significant abundance. The right spectrum clearly demonstrates that this discharge setup is capable of providing VUV radiation far below 100 nm. The broad emission around 65 nm is

4 Results and Discussion

assigned to strong Ar II (ionized argon) and N II transitions. The remaining part of the spectrum is again dominated by N I, O I and H I emissions. It is stressed that these results were obtained under operating conditions that are not directly compatible with APPI-MS. First, the impact of the helium background gas on the discharge was not known and second, subsequent discharge chemistry could have been more extensive, since the distance between grating and discharge region was about 200 mm, whereas the distance to the sample gas flow within the transfer capillary is less than 1 mm. The large distance in the spectroscopic setup is a reasonable explanation for the observed extensive impurity emissions. These issues will be resolved in future experiments. Nevertheless, both spectra give a fairly good impression of (i) the ability of the lamp to produce significant amounts of VUV radiation below 100 nm, (ii) the advantage of the windowless lamp design, and (iii) the potential of this type of discharge lamp to function as a selectable VUV radiation source by using an appropriate discharge gas mixture. Furthermore, the VUV emission measurements with the commercially available APPI lamp will be repeated using a windowless spectrometer.

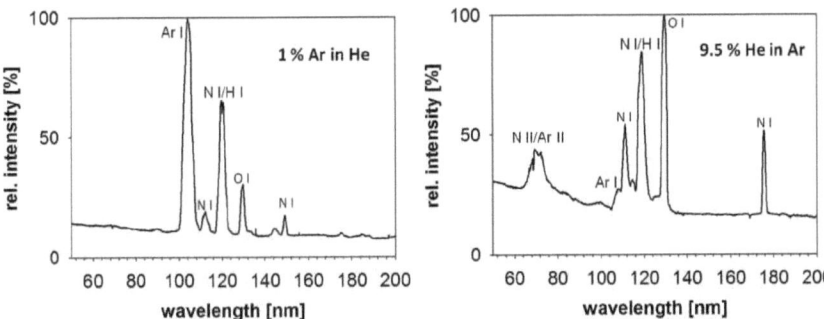

Figure 38: VUV measurement of a spark discharge lamp below 105 nm (in cooperation with Resonance Ltd. Barrie, On, Canada). The background gas of the spectrometer was helium at a pressure of 830 mbar. The gas composition delivered through the electrodes was (left) 1 % argon in helium and (right) 9.5 % helium in argon.

4.2.3.4 Impact of Different Ionization Positions on MS Spectra

So far experiments were centered on the development and characterization of miniature spark discharge lamps for implementation in MS transfer capillaries. The goal was to significantly reduce ion transformation processes that are dependent on the total number of collisions, and at the same time provide high neutral analyte densities for efficient ion

4 Results and Discussion

generation. Based on degradation product studies, the following two sections will give an impression of the effects when shifting the ionization position further downstream along the capillary main axis.

a) Impact on negative ion mode

In Figure 39 (a,b) the effect of reduced ion transit times within the capillary is demonstrated using the negative ion mode. A degradation experiment with p-xylene was repeated three times at three different ionization positions: (i) Within the laminar flow ion source (cf. Figure 12, 8), (ii) on position 15 cm, and (iii) on position 17.5 cm of the transfer capillary. The different transit times thus accounted for 5 ms, 0.13 ms and 0.02 ms, respectively (cf. chapter *4.2.2.3 LFIS – APPI (b)* and Table 2). Figure 39 (a) shows the recorded mass spectra at the beginning of the experiment, with all chemicals (1 ppmV p-xylene, 5 ppmV MeONO, 3 ppmV NO) injected into the photoreactor. The most abundant peak in each spectrum is assigned to the *[M-H]⁻* of HNO_3 ($m/z = 62$) always present due to the formation as a byproduct in a degradation run (see also reaction 9, page 28)

$$HO + NO_2 + M \rightarrow HNO_3 + M \qquad \text{(rxn 14)}$$

with a termolecular rate constant of $k = 2.5 \cdot 10^{-30}$ $cm^6 \cdot mlecule^{-2} \cdot s^{-1}$ [138]. Figure 39 (b) shows mass spectra after 25 min of the degradation experiment for comparision. The spectrum generated with the ionization position in the laminar-flow ion source (5 ms) remained virtually unchanged. In contrast, the spectra obtained from ionization positions within the transfer capillary show significantly more signals that are assigned to oxygenated products of the degraded p-xylene. The extensive loss of MS information along the 5 ms transit time is assigned to secondary ion transformation processes of the types:

$$[M]^-/[M-H]^- + HNO_3 \rightarrow [MH]/[M] + NO_3^- \qquad \text{(rxn 15)}$$

$$[M]^-/[M-H]^- + [P]^+/[P+H]^+ \rightarrow [M]/[MH] + [P] \qquad \text{(rxn 16)}$$

$$[M]^-/[M-H]^- + [P]^+/[P+H]^+ \rightarrow [MP]/[MPH] \qquad \text{(rxn 17)}$$

with *[M]* and *[P]* being any chemical species present in the matrix. In reaction 14 the species with the lower proton affinity (HNO_3) is favored and reactions 15 - 17 consider charge neutralization. In other words, the ion signal distribution is thermodynamically controlled within 5 ms and mass spectrometric information reflecting the true neutral composition is lost,

4 Results and Discussion

whereas ionizing within the transfer capillary shifts the ion signal distribution towards kinetic control, hence more MS information is preserved. These reactions will be discussed in more detail in chapter *4.3.3.2 APPI/APLI-Negative Chemical Ionization (NICI)*.

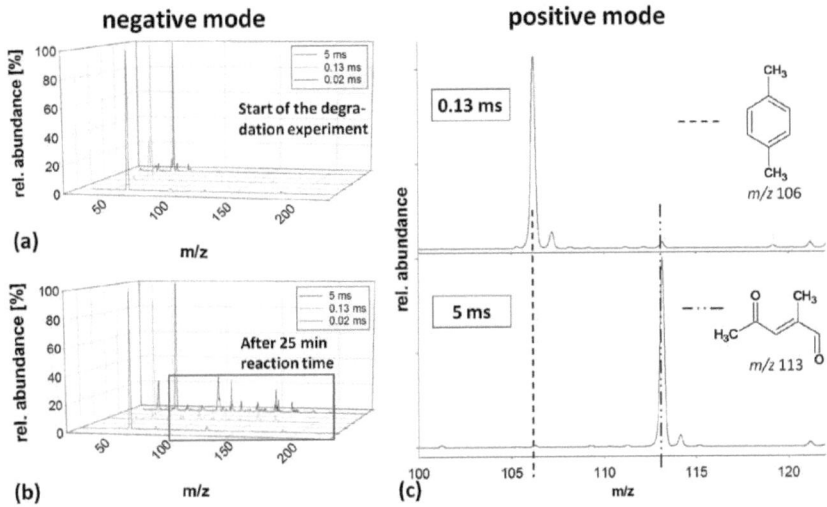

Figure 39: Mass spectra obtained at different ionization positions. The positions are denoted as 5 ms, 0.13 ms and 0.02 ms referring to the transit time between the ionization position and the capillary exit. (a) Comparison of negative ion mass spectra at the starting point of a degradation experiment with p-xylene (1 ppmV), MeONO (5 ppmV) and NO (3 ppmV) in synthetic air present, (b) comparison of negative ion mass spectra of the same degradation experiment after 25 min of reaction time. (c) Comparison of positive ion mass spectra recorded at the 5 ms and 0.13 ms position, with identical p-xylene degradation composition present.

b) Impact on positive ion mode

The impact on the positive ion mode upon changing the ionization position is demonstrated in Figure 39 (c). Both spectra were recorded at the same reaction time of one p-xylene degradation experiment. The spectrum recorded with ionization on the 5 ms position leads to the conclusion that p-xylene (m/z 106) has been quantitatively degraded. Consequently, the main peak appears at m/z 113, assigned to the $[M+H]^+$ of the unsaturated 1,4-dicarbonyl, which is one of the expected major degradation products [6]. However, the mass spectrum recorded at the 0.13 ms ionization position reveals that m/z 106 for p-xylene is

4 Results and Discussion

the most abundant signal. This effect is explained by the considerable proton donor characteristic of the ionized p-xylene, leading to the following ion transformation process:

$$[M]^+ + [P] \rightarrow [M-H] + [P+H]^+ \qquad \text{(rxn 18)}$$

with *[M]* as p-xylene and *[P]* as the 1,4-dicarbonyl. Hence, the p-xylene acts as a proton donor for its own degradation product. It follows that with sufficient time after the ionization step the relative ion distribution does not reflect the neutral composition of the sample anymore. Consequently, ionization on the 5 ms position suggested a much faster progress of the degradation reactions than in fact it was. This renders kinetic studies of degradation experiments, where rate constants ought to be investigated, nearly impossible with common API-MS methods. However, at the 0.13 ms position the molecular ion of the primary photo ionized 1,4-dicarbonyl became visible at m/z 112. The impact of superimposed ion chemistry on this signal is believed to be negligible since the concentration of other species, where proton transfer could play a role, are rather low. Additionally, other neutral degradation products would readily be titrated away by the presence of high concentrations of the p-xylene radical cation. Consequently, with the introduction of the novel APPI approach at least rough quantitative estimates[5] concerning the relative concentrations of neutral degradation products within the sample became possible. This discussion will be extended in chapter *4.3.3 ITP via Chemical Ionization*.

It was shown that within the operating conditions of the present mass spectrometer the lowest possible dwell time of generated ions at elevated pressure in fact considerably reduces ion transformation processes. APPI on transfer capillaries in common mass spectrometers preserves significantly more MS information; thus the recorded ion signal distribution much closer reflects the neutral composition. At this point some signal fluctuations at the 20 µs position, especially in the alternating mode of the mass spectrometer, should be mentioned. This problem has not been completely understood, however, it is tentatively assigned to the impact of the discharge on the potential applied between the end of the capillary and the skimmer. It is believed that with the implementation of lamp design 3 at this position (lower breakdown voltage; use of the DD20_10 C-Lader as the power supply) the observed effects are significantly reduced. Hence, most degradation studies have not been performed with

---

[5] Mass discrimination effects due to the operating conditions of the instrument and individual VUV absorption cross sections still interfere for true quantification. However, the variation in the latter is not as significant and consequently not as selective as, e.g. in APLI.

ionization on the position with the lowest possible transit time, but on the 0.13 ms position, which still gave reasonable results in terms of ITP reduction.

4.3 Ion Transformation Processes (ITP)

So far the discussion might lead to the conclusion that ion transformation processes are *per se* undesired and need to be completely eliminated. As already mentioned in the introduction, this is not the case. ITPs bear the ability to provide valuable structural information and they are capable of significantly enhancing the mass spectrometric sensitivity of certain species. However, and this is the pivotal point, in order to correctly interpret a mass spectrum, it is necessary to know what kind of transformation processes between the ionization step and the detection step occurred and also to which extend. This generally complicates the mass spectrometric analysis of unknown samples investigated with API-MS, since in lieu of a reference spectrum that illustrates an ITP reduced ion signal distribution, these processes are difficult to determine. Four types of transformation processes that were of fundamental relevance within this work of mass spectrometric analysis of atmospheric degradation product studies will be dealt with in the following chapters.

4.3.1 Unintended Collision Induced ITP

Fragmentation processes in mass spectrometry may be induced or enhanced by increasing the kinetic energy of an ion with potential gradients along the travel path. This results in higher collision energies, which might eventually be sufficient to overcome the potential activation barrier and thus induce fragmentation. The esquire6000 exhibits two acceleration stages that should be considered, namely (i) the transfer region from the capillary exit into the high vacuum section (cf. Figure 4, right) and (ii) the "trap drive" region governed by parameters that determine the potential well in which ions are being captured within the trap (cf. *3.1 Mass Spectrometer*). The first case was of significance with one of the early 20 µs-position lamp designs in which the discharge was still physically separated through a LiF window. The efficient guidance of ions into the high vacuum stage required optimized voltages on the capillary exit and the skimmer of 360 V and 0 V, respectively. With this maximum settable potential gradient severe fragmentation patterns of analytes were observed.

4 Results and Discussion

It is noted that commonly applied voltage differences range between 70 V and 100 V. At this time it is not known why the transmission efficiency of this early design required such high voltages. Using the same capillary, but a further upstream ionization position (laser or 5 ms position), the common voltages were applied for sufficient transfer, so this effect was unambiguously assigned to the operating conditions of this lamp type. Furthermore, with the introduction of the windowless approach on this outermost position this effect was not present anymore, merely the mentioned lack of stability in the alternating mode was observed. The second case, concerning the trap drive parameter, turned out to be a balancing act between optimized trapping efficiency and induced fragmentation processes. The applied voltages force the ions into a potential well, in other words they are accelerating. At this point the presence of helium as the buffer gas is supposed to cause collisional cooling effects; however, with sufficient gain of kinetic energy unintended collision induced dissociation becomes obvious. Figure 40 demonstrates the impact of this type of ITP.

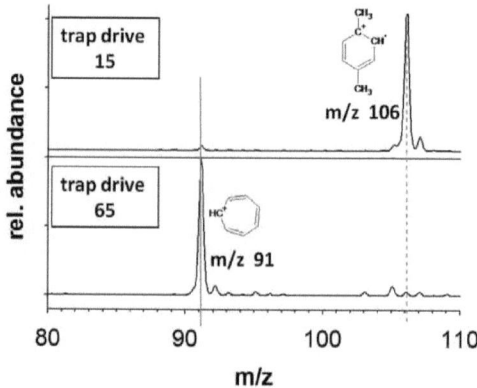

Figure 40: Impact of the "trap drive" parameter on induced fragmentation of ionized p-xylene.

A sample of around 1 ppmV of p-xylene in synthetic air was delivered from the photoreactor and irradiated with the DPSS laser in coaxial configuration of the laminar-flow ion source. The only parameter changed was the applied trap drive, with a fixed value of "15" in the upper and "65" in the lower spectrum. These correspond to a small and a fairly high (but still

4 Results and Discussion

in the commonly applied range) accelerating potential, respectively. The induced fragmentation process with the higher trap drive value leads to virtually complete loss of the parent molecule at m/z 106 in favor of the tropylium cation at m/z 91 according to

$$[C_8H_{10}]^+ + E_{kin} + He \rightarrow [C_7H_7]^+ + CH_3 + He \qquad \text{(rxn 19)}$$

In order to exclude interfering mass discrimination effects (cf. *3.1 Mass Spectrometer, b*) between m/z 91 and m/z 106, due to changes in the trapping efficiency, toluene (m/z 92) was added with a concentration of around 2 ppmV into the photoreactor. Once more the values "15" and "65" were applied as the trap drive level and again the signal of the tropylium cation and the p-xylene radical cation switched intensities, whereas the relative toluene signal was virtually unaffected. Consequently, the observed abundance of the tropylium cation at higher trap drive levels was unambiguously assigned to unintended CID.

4.3.2 Neutral Radical Induced ITP (NRITP) [82][6]

Upon vaporizing a 100 µL·min⁻¹ flow of a 10 µM ACN or methanol solution of pyrene using the Apollo™ source (Figure 7), and switching on either the 193 nm laser or the commercially available Kr-RF lamp, three signals in addition to the radical cation of pyrene (*[M]⁺*, m/z = 202) at m/z = 217, 218, and 219, i.e., *[M+15]⁺*, *[M+16]⁺*, and *[M+17]⁺* were detected. The chemical species leading to these signals are referred to as "oxygenated products" of the primary generated radical cation. Identical results were obtained when gaseous pyrene was delivered directly at lower ppbV mixing ratios, i.e., in the absence of solvents, but in the presence of around 2.4 % oxygen and ~100 ppmV water, as determined in *4.1.3 H₂O and O₂ Background Concentrations [82]*.

This experimental observation is pivotal to the present work, particularly with respect to the investigation of oxidation pathways of aromatic hydrocarbon cations in API-MS using ionizing radiation below 200 nm. The speculated key radicals and atoms potentially responsible for the appearance of artificially produced oxygenated *ions* are O(³P), OH, H, and Cl, generated upon photolysis of O₂, H₂O, and chlorinated molecules, such as the solvents CH₂Cl₂, CHCl₃ and CCl₄, respectively (cf. *1.2.1.2 Atmospheric Pressure Photo Ionization (APPI)*). Each of these species potentially drives a comprehensive chemistry, which can be

---

[6] It is noted that most of the content of this chapter has already been published in 2009 [82].

4 Results and Discussion

described using well known atmospheric processes. Table 4 summarizes the most relevant reactions that have to be considered.

Table 4: Neutral radical reactions considered in the present investigation.

No	Reaction	Rate constant k*	Absorption cross section σ [cm^2 molecule^{-1}] 193 nm/124 nm**
1	$O_2 + h\nu \rightarrow 2\ O(^3P)$	22.5 $^+$	$3 \cdot 10^{-22}/\sim 10^{-20}$
2	$H_2O + h\nu \rightarrow H + OH$	135 $^+$	$1.8 \cdot 10^{-21}/\sim 10^{-17}$
3	$CH_nCl_{4-n}\ (n=0-2) + h\nu \rightarrow CH_nCl_{4-n-1(2)} + Cl$	$6.8 \cdot 10^{4\ +}$	$9 \cdot 10^{-19}/\sim 10^{-17}$
4	$MeOH + h\nu \rightarrow MeO + H$	$7.5 \cdot 10^{3\ +}$	$10^{-19}/\sim 10^{-17}$
5	$O_3 + h\nu \rightarrow O(^1D) + O_2$	$3 \cdot 10^{5\ +}$	$4 \cdot 10^{-19}/\sim 10^{-17}$
6	$O_2 + O(^3P) + N_2 \rightarrow O_3 + N_2$	$6.0 \cdot 10^{-34\ +++}$	n.a.
7	$O(^1D) + N_2 \rightarrow O(^3P)$	$2.6 \cdot 10^{-11\ ++}$	n.a.
8	$O(^1D) + H_2O \rightarrow 2\ OH$	$2.2 \cdot 10^{-10\ ++}$	n.a.
9	$Cl + O_3 \rightarrow ClO + O_2$	$1.2 \cdot 10^{-11\ ++}$	n.a.
10	$Cl + O_2 + N_2 \rightarrow ClOO + N_2$	$2.7 \cdot 10^{-33\ +++}$	n.a.
11	$ClO + O(^3P) \rightarrow O_2 + Cl$	$3.8 \cdot 10^{-11\ ++}$	n.a.
12	$ClOO + Cl \rightarrow 2\ ClO$	$1.2 \cdot 10^{-11\ ++}$	n.a.
13	$ClOO + N_2 \rightarrow Cl + O_2 + N_2$	$6.2 \cdot 10^{-13\ ++}$	n.a.
14	$ClOO + O(^3P) \rightarrow O_2 + ClO$	$5.0 \cdot 10^{-11\ ++}$	n.a.
15	$O_3 + H \rightarrow HO + O_2$	$2.9 \cdot 10^{-11\ ++}$	n.a.

* T = 298.15 K; first order, second order and third order rate constants are given in units of [s^{-1}], [cm$^3 \cdot$molecule$^{-1} \cdot$s^{-1}], and [cm$^6 \cdot$molecule$^{-2} \cdot$s^{-1}], and are denoted as $^+$, $^{++}$ and $^{+++}$, respectively. Photolysis rate constants are calculated assuming a 10 ns laser pulse of 1 mJ energy at λ = 193 nm and an irradiated area of 1 cm^2. This corresponds to a photon density of 7.5$\cdot$10^{22} photon$\cdot$cm$^{-2} \cdot$s^{-1}. The calculation of the rate constant for the PI lamp is not possible, due to the unknown photon density in the present set-up.
** Rate constants are taken from [138] and cross sectional data are taken from [52-57,139]

Since common atmospheric degradation product studies are performed in synthetic air with about 20 % oxygen present and furthermore a considerable concentration of water is build up during such an experiment, the necessity of investigating neutral radical induced ITPs became soon apparent. In addition, it is recalled that the expected *neutral* degradation products are of oxygenated nature as well, which might significantly interfere in MS data interpretation. In the following a thorough study, which is exemplarily based on the oxidation pathway of the

4 Results and Discussion

pyrene radical cation, will be given. The results are discussed in the context of the novel APPI approach and the typical conditions prevailing in atmospheric degradation studies.

For appropriate experiments a flow-tube assembly was constructed, which allowed the spatial separation of the neutral radical production region and the ionization region (cf. *3.1.3 Setup for Neutral Radical Induced ITP Studies [82]* and Figure 11).

4.3.2.1 Evidence for Ion-Neutral Radical Chemistry

To demonstrate that the proposed neutral radical ion chemistry was responsible for the observed additional signals, the following experiment with the flow-tube assembly was performed. First, as shown in Figure 41 (top, a), the 248 nm laser generated a strong pyrene radical cation $[M]^+$ signal recorded at *m/z* 202.

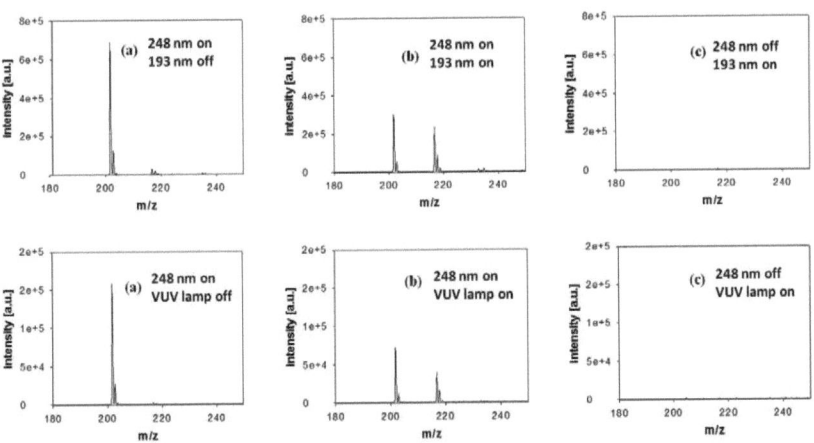

Figure 41: Mass spectra recorded upon irradiation of a nitrogen carrier gas flow in the presence of ~ 6 ppbV pyrene, ~170 ppmV CCl$_4$, ~100 ppmV O$_2$, and ~ 200 ppmV H$_2$O. Top: (a) 248 nm laser on and 193 nm laser off; (b) both lasers on; (c) 248 nm laser off, 193 nm laser on. Bottom: (a) 248 nm laser on and VUV lamp off; (b) 248 nm laser on, VUV lamp on; (c) 248 nm laser off, VUV lamp on.

In a second experiment the 193 nm laser beam was switched on, resulting in a mass spectrum with the additional signals of $[M+15]^+$, $[M+16]^+$ and $[M+17]^+$ (Figure 41, top, b), while the sum of the signals at *m/z* = 202, 217, 218, 219 remained virtually constant. Hence, the signal

4 Results and Discussion

intensity of m/z 202 decreased. Turning only the 248 nm laser off resulted in a nearly blank mass spectrum (Figure 41, top, c). This experiment clearly demonstrated: (i) The flow-tube setup allowed the spatial separation of ion and neutral radical production; (ii) neutral radical formation was mostly absent at a wavelength of 248 nm, and (iii) the additional signals generated with the 193 nm light were due to secondary products resulting from the reaction of *neutral* radicals with the radical cation. When the 193 nm laser was replaced with the Kr-RF lamp almost identical datasets were obtained (cf. Figure 41, bottom, a-c). It was thus concluded that upon APPI at 124 nm a rich neutral radical chemistry is initiated, which has a significant impact on the signal distribution in the resulting mass spectra.

The relative and absolute intensities of the mass signals of the oxidation products were strongly changed by variation of the O_2, H_2O and CH_nCl_{4-n} (n=0–2) concentrations, respectively, while the sum of the mass spectrometric signals remained constant. Changing the ionizing laser position along the quartz tube main axis, i.e., changing the reaction time, also strongly affected the mass signal distribution up to the complete loss of $[M]^+$. Collisionally induced dissociation of any of the three additional species resulted in loss of m/z = 28 and 18, assigned to CO and H_2O, respectively. This strongly suggested the addition of an oxygen atom to the radical cation in some way, which was used as justification of denoting the additional signals as "oxygenated" products.

4.3.2.2 Oxidation of the Pyrene Radical Cation - Feasible Pathways

For simplification of the mechanistic studies, only the 193 nm laser in the flow tube setup (cf. Figure 11) was used as photolysis *and* at the same time ionization source and positioned between the exit of the glass tube and the entrance of the transfer capillary. With a carrier gas velocity of 46 cm·s^{-1}, a laser pulse repetition rate of 10 Hz, and a laser beam width of 1 cm within the flow direction, a spatially well defined volume ("package") was irradiated by the laser light only once. This ensured that (i) the radical cation generation was the initial process of all subsequent reactions and (ii) generated ozone was not photolysed yielding $O(^1D)$-atoms (cf. Table 4, reaction No 5), which would have rendered the further data interpretation much more difficult. In addition, the kinetic simulations were simplified, as specific neutral and ionic radical concentrations were generated as a "package" from the temporally and spatially well defined single laser pulse. The package traveled toward the sampling orifice of the mass spectrometer. It is pointed out that the selected reaction time,

4 Results and Discussion

defined by the laser position, and the initial concentrations of reactive reagents, defined by the total photon flux of one single laser pulse, determined the relative and absolute $[M]^+$, $[M+15]^+$, $[M+16]^+$ and $[M+17]^+$ signal intensities.

Experimental and theoretical investigations lead to two possible reaction channels following the ionization step of the pyrene radical cation. One favors the direct addition of neutral radicals, such as OH and O(^{3}P) to the aromatic ring system. The speculated second path is initiated by H-abstraction from the pyrene skeleton and subsequent addition of OH, O(^{3}P), or even closed shell molecules, such as H_2O, to form the oxygenated products. These two cases are discussed in the following; all molecular structures referenced with a "#" are shown in Figure 42, all reactions referenced with a "No" are summarized in Table 5.

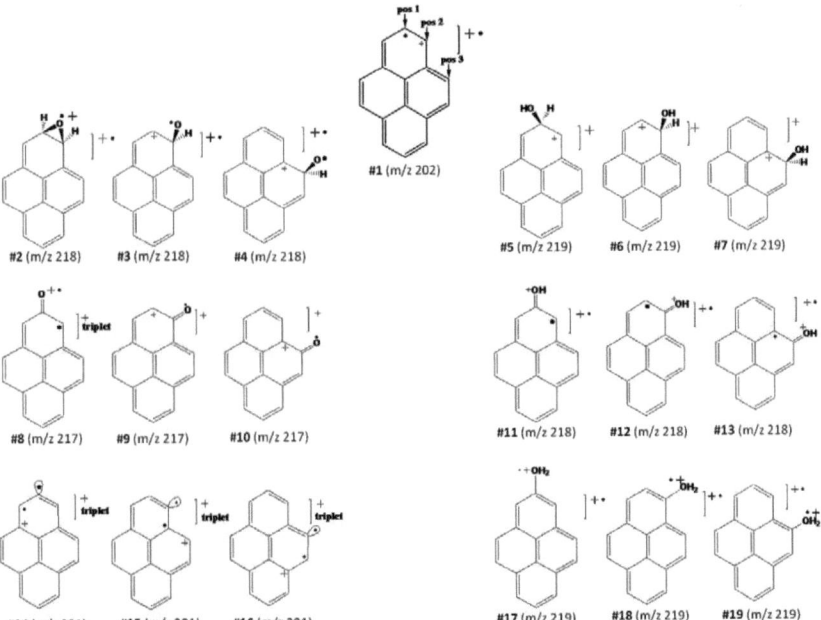

Figure 42; Possible structures within the oxidation pathways of the pyrene radical cation.

4 Results and Discussion

Table 5: Computated ΔG_R values [eV] for the examined reactions.

No	Reactant structure of the pyrene system*	Other reagents	Product structure of the pyrene system*	Other products	Gibbs free energy of reaction ΔG_R [eV]**
1	#1	O(^{3}P)	#2/ #3/ #4	---	- 2.00/ -1.58/ -1.09
2	#1	OH	#5/ #6/ #7	---	-0.78/ -1.52/ -1.04
3	#2/ #3/ #4	---	#8/ #9/ #10	H	1.75/ 0.44/ 0.40
4	#5/ #6/ #7	---	#11/ #12/ #13	H	1.14/ 1.69/ 1.22
5	#5/ #6/ #7	---	#2/ #3/ #4	H	2.94/ 4.09/ 4.10
6	#11/ #12/ #13	---	#8/ #9/ #10	H	3.56/ 2.84/ 3.28
7	#1	---	#14/ #15/ #16	H	4.57/ 4.48/ 4.51
8	#14/ #15/ #16	O(^{3}P)	#8/ #9/ #10	---	-4.82/ -5.62/ -5.21
9	#14/ #15/ #16	OH	#11/ #12/ #13	---	-4.22/ -4.30/ -4.33
10	#14/ #15/ #16	H$_2$O	#17/ #18/ #19	---	-0.71/ -0.57/ -0.56
11	#17/ #18/ #19	---	#11/ #12/ #13	H	1.13/ 0.91/ 0.87
12	---	H + OH	---	H$_2$O	-4.64
13	---	H + O(^{3}P)	---	OH	-4.16
14	---	H + H	---	H$_2$	-4.24
15	---	H + O$_3$	---	OH + O$_2$	-3.91
16	---	H + O$_2$	---	HO$_2$	-1.84
17	---	H + Cl	---	HCl	-4.03
18	---	H + ClO	---	HOCl	-3.69
19	---	H + ClOO	---	ClO + OH	-1.67

* pyrene structures are numerated as shown in Figure 42; ** in the case of possible singlet and triplet structures only the most stable one is listed, as denoted in Figure 43.

a) Oxidation via direct addition of O(^{3}P) and OH

Pyrene is a molecule of high symmetry, belonging to the point group C1. The three different possible positions of an electrophilic attack are shown in structure #1. The addition of neutral species to the pyrene radical cation is not favorable for closed shell molecules, e.g., O$_2$ or H$_2$O, as reported earlier by Le Page et al. [70]. These results were fully confirmed within this work, since with 248 nm excitation virtually no other ions than $[M]^+$, even in the presence of large quantities H$_2$O and O$_2$, were detected. However, the same authors showed that the addition of neutral radicals such as O(^{3}P) or H-atoms to the ionized ring system are fast reactions with bimolecular rate constants on the order of $k \approx 10^{-11} - 10^{-10}$ cm$^3 \cdot$ molecule$^{-1} \cdot$ s^{-1}. Wood et al. [140] also observed the rapid OH-addition to ionized C60 fullerene, another

4 Results and Discussion

highly condensed aromatic ring system. The performed DFT calculations fully supported this reaction pathway as well (cf. Figure 43, left). The addition reactions of O(^{3}P) or OH to the pyrene radical cation (reaction No 1 and No 2) leading to structures #2 - #4 and #5 - #7, were calculated to be exergonic by ΔG_R = -2.00 to -1.09 eV and -1.52 to -0.78 eV, respectively. The subsequent step in this reaction path is speculated to be H-atom abstraction from the oxygenated pyrene, e.g., by OH-radicals. OH-radicals are always present in APPI due to photolysis of H_2O. For structures #2 - #4 transformed to structures #8 - #10, Gibbs free energies range from ΔG_R = -4.24 to -2.89 eV (No 3 combined with No 12). So far, the analogy to the well described nucleophilic aromatic substitution reaction type is striking. In the cases of structures #5 - #7 two H-atom abstraction channels by OH-radicals are feasible (reaction No 4/No 5 combined with, e.g. No 12). First, according to the nucleophilic aromatic substitution scheme, removal of the H-atom from the carbon skeleton leads to structures #11 - #13 (ΔG_R = -3.50 to -2.95 eV). The second channel is the H-atom abstraction from the hydroxyl group, yielding the same products as with O(^{3}P)-addition to the pyrene radical cation (#2 - #4; ΔG_R = -1.70 eV to -0.54 eV). The latter compounds can repeatedly react with neutral radicals, as shown before, to yield products with structures #8 - #10. The products from reaction No 4 (#11 - #13) can further react with neutral radicals via H-atom abstraction from the hydroxyl group to form the products #8 - #10 (ΔG_R = -1.80 eV to -1.08 eV; No 6 combined with No 12).

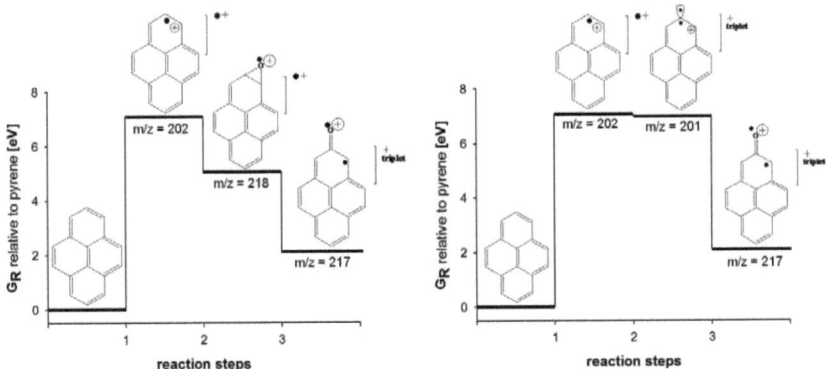

Figure 43: (left) Reaction steps: (1) [M] → [M]$^+$, (2) [M]$^+$ + O → [M+O]$^+$, (3) [M+O]$^+$ + OH → [M-H+O]$^+$ + H_2O (right) Reaction steps: (1) [M] → [M]$^+$, (2) [M]$^+$ + OH → [M-H]$^+$ + H_2O, (3) [M-H]$^+$ + O → [M-H+O]$^+$. ΔG_R = G_R relative to pyrene (g).

4 Results and Discussion

Obviously, all intermediates are eventually converted to the most stable species with m/z 217. This trend is very well supported by experimental data. The signal at m/z 217 increases with increasing neutral radical concentration and/or with increasing reaction time. It is stressed that the neutral radical concentrations are exceeding the parent ion concentration by orders of magnitude.

b) Oxidation via the pyrenyl cation [M-H]$^+$

The primary step in this second case is the abstraction of an H-atom from the carbon skeleton of the radical cation to form *[M-H]$^+$* (cf. Figure 43, right). However, the C-H bond dissociation enthalpy is rather high, experimentally determined to be 4.6 eV [141,142] and calculated to be $\Delta H_{d(C-H)}$ = 4.93 eV, 4.84 eV and 4.88 eV for the positions 1, 2 and 3, respectively. Thus, reactions leading to the pyrenyl cation (#14 - #16) in its triplet state as the most stable conformation are rather limited.

In first instance it was speculated that the available maximum excess energy ($E_{excess\,max}$), i.e., the difference of the total energy of the two absorbed photons (E_{total}) and the ionization energy (E_i = 7.41 eV [143]), with an assumed hypothetical electron kinetic energy of E_{kin} = 0 eV,

$$E_{excess\,max} = E_{total} - E_i \qquad (\text{eq } 34)$$

could be sufficient to initiate uni-molecular fragmentation reactions of the types

$$[M]^+ \rightarrow [M\text{-}H]^+ + H \qquad (\text{rxn } 20)$$

$$[M]^+ \rightarrow [M\text{-}2H]^+ + 2H/H_2 \qquad (\text{rxn } 21).$$

Following two-photon ionization at 248 nm ($E_{excess\,max}$ = 2.6 eV) this pathway was immediately excluded. Absorption of two 193 nm photons contributes an excess energy of 5.4 eV to the pyrene radical cation, which appeared to be sufficiently high. However, Ling et al. determined the appearance energy (AE) of the pyrenyl cation *[M-H]$^+$* with time-resolved photo-ionization mass spectrometry (TPIS) to be 15.2 eV [142]. Hence, an effective excess energy of at least 7.8 eV is required for the formation of the pyrenyl cation and thus uni-molecular decomposition after 193 nm (1+1) REMPI was excluded as well. More importantly, it is highly unlikely that the parent ion is generated by direct two-photon absorption. As discussed below, ultra-fast relaxation of the intermediately pumped S_4 state

4 Results and Discussion

into the S_1 manifold strongly suggests that the parent ion is carrying an excess energy far below the nominal $(6.3 + 6.3 - E_i)$ eV.

Among the various neutral radical species typically generated in the ion source, only the hydroxyl radical seems to be able to abstract an H-atom of the pyrene radical cation carbon skeleton. Enthalpy release and the Gibbs free energy for the dissociation of the H-OH bond accounts for $\Delta H_{d\,(H\text{-}OH)} = 4.96$ eV (literature value: 5.11 eV [144]) and $\Delta G_R = 4.64$ eV (literature value: 4.83 eV [138]), respectively. The Gibbs free energy of the C-H dissociation reaction No 7 is calculated to be in the range of $\Delta G_R = 4.48$ eV and 4.57 eV. Once the pyrenyl cation is formed it shows high reactivity towards neutral radical species and even closed shell molecules such as H_2O [70, 141, 145]. The addition of $O(^3P)$ and OH (No 8/No 9) leads to structures #8 - #10 and #11 - #13 with ΔG_R between -5.62 eV and -4.82 eV and ΔG_R in the range of -4.33 eV and -4.22 eV, respectively (cf. Figure 43, right). Henceforward, the sequence proceeds via reaction No 6 as described in the preceding section of the direct addition channel. The addition of H_2O to the pyrenyl cation forms structures #17 - #18 with ΔG_R between -0.71 eV and -0.56 eV. The subsequent removal of the hydroxyl H-atoms leads to the most stable phenoxy type species #8 - #10 ($m/z = 217$; No 11/No 6), again in analogy to the direct addition channel. Another possible route is the addition of molecular oxygen to the pyrenyl cation, leading to a peroxy radical ($m/z = 233$). This mass was recorded as minor signal (~0.002 % rel. abundance), when 420 ppmV H_2O and 0.7 % O_2 were present. However, this signal could also be attributed to the two-fold OH or $O(^3P)$ addition with subsequent H-atom abstraction. If the reaction proceeds via peroxy radical intermediates then the corresponding signal should be rather low, as observed: (i) Very small peroxy radical population is formed, due to the comparably low third order rate constant(s) or simply due to the missing pyrenyl cation as a precursor, (ii) the subsequent reaction of forming a hydroperoxy radical, followed by O-OH bond cleavage in order to release an OH-radical and the most stable phenoxy type species #8 - #10 would lead to a rapid degradation of the peroxy radical cation concentration.

Consequently, also the pyrenyl cation reaction channel favors phenoxy-type end-products (m/z 217) with increasing radical concentration and/or with increasing reaction time.

Though $[M\text{-}H]^+$ species are far more reactive than the corresponding $[M]^+$ precursors, the major reaction pathway is attributed to the direct addition of OH and $O(^3P)$. This is rationalized in the following manner: (i) Generation of pyrenyl cations is unlikely, mainly due

4 Results and Discussion

to the high dissociation enthalpy of the C-H bond, (ii) the corresponding reaction of the neutral atmospheric degradation of aromatic compounds with OH proceeds up to 90 % via the OH-addition channel [1], and (iii) the signal recorded at $m/z = 201$, attributed to $[M-H]^+$, has a very low abundance. The latter argument is of course also in accord with the high reactivity of $[M-H]^+$.

As conclusion, the OH/O(^{3}P) radical addition channel to the charged aromatic ring system is much more likely to be responsible for the efficient oxidation of primarily formed radical cations. In the following section on kinetic simulations, the oxidation pathway via the pyrenyl cation is thus not considered further.

4.3.2.3 Oxidation of the Pyrene Radical Cation - Kinetic Investigations

Further elucidation of possible oxidation mechanisms was carried out by comparison of experimentally recorded mass spectra with results from kinetic simulations. In these experiments, the 193 nm laser was running with a repetition rate of 10 Hz (see above) and positioned 0.5 to 1.5 cm upstream of the entrance of the MS inlet capillary (cf. Figure 11). Considering the flow velocity of 46 cm·s^{-1}, the average reaction time for a generated "package" up to the entrance of the MS accounted for 21 ms. The total reaction time in addition with the calculated 2 ms transit through the transfer capillary (cf. Table 2) thus amounted to 23 ms. The abundances of major signals in the recorded mass spectrum are summarized in Table 6. The summed peak areas of the oxygenated products account for 92% of the $[M]^+$ peak area. Assuming that the addition channel is dominating, O(^{3}P) and OH attack could potentially lead to primary oxygenated products with $m/z = 218$ and 219, respectively.

Table 6: Major peak abundances in a mass spectrum recorded upon irradiating a flow of N$_2$ at 1 bar with 420 ppmV H$_2$O, 0.7 % O$_2$ and 6 ppbV pyrene present.

m/z	assigned structures*	rel. peak area [%]**
202	#1	100
217	#8 - #10	70.4
218	#2 - #4 and #11 - #13	18.0
219	#5 - #7	3.7

* cf. Figure 42; ** Relative to the pyrene radical cation $[M]^+$.

4 Results and Discussion

Computational simulation of the resulting charged product distribution recorded after these 23 ms of reaction time required an estimate of the initial pyrene radical cation concentration according to equations 12 and 13 (see page 48). The excitation of pyrene at 193 nm results in initial population of the S_4 state which relaxes to the S_1 manifold on the ps time-scale [146]. The S_1 state has an estimated lifetime of $\tau_{res} \approx 100$ ns [146-151], which is 10 times larger than the laser pulse width. Thus, the transition into the ionization region originates from the S_1 manifold. The overall absorption cross section σ is to a good approximation the product of the subsequent single step absorption cross sections $\sigma_{1,2} = 1.7 \cdot 10^{-17}$ cm^2 [44] and the lifetime of the resonant state τ_{res}

$$\sigma = \sigma_1 \cdot \sigma_2 \cdot \tau_{res} \qquad (eq\ 35).$$

The quantum yields for both absorption steps are assumed to be $\varphi_1 = \varphi_2 = 1$ molecule·photon^{-1}, the photon flux is $\Phi = 6.0 \cdot 10^{22}$ photon s^{-1} cm^{-2} and the laser pulse duration is $t_{pulse} = 10$ ns. Therefore, an initial pyrene concentration of 6 ppbV leads to a calculated pyrene radical cation concentration of $\{[M]^+\} \approx 1.5 \cdot 10^8$ molecule·cm^{-3} (about 5 pptV) generated within one laser pulse. Since the neutral radical concentrations are at least one order of magnitude higher, all reactions proceed in pseudo first order with respect to the charged species. It follows that the product peak abundance distribution relative to the abundance of $[M]^+$ is independent of the initial pyrene radical cation concentration and is controlled by the concentration of the neutral radicals.

a) Impact of O_2, $O(^3P)$, and O_3 on the ion distribution

The initial $O(^3P)$ mixing ratio is calculated as 3.1 ppbV when 0.7 % oxygen is present in the source enclosure. Applying a rate constant of $k = 9.5 \cdot 10^{-11}$ cm^3·molecule^{-1} s^{-1} [70] for the reaction of $O(^3P)$ with pyrene radical cations, 16 % of the latter are converted into the oxygen adduct (#2 - #4) within 23 ms reaction time. Hence, the hypothetical outcome of the mass spectrum would exhibit two signals, $[M]^+$ and $[M+16]^+$, with a ratio of about 100 : 19. However, due to the high mixing ratio of O_2 the competing ozone formation reaction (No 6, Table 4) is proceeding much faster, nearly quantitatively converting the entire $O(^3P)$ population into O_3. Ozone, however, is a rather inert species in the present environment towards ionized aromatic species, as shown by Mendes et al. [152]. Since the laser pulse irradiates the spatially defined "package" only once, photolysis of ozone (No 5, Table 4) and thus formation of $O(^1D)$ is excluded. Taking the ozone generation reaction into account, only

4 Results and Discussion

0.3 % of the initial pyrene radical cations would be converted to the oxygen adducts #2 - #4 within 23 ms reaction time and two peaks with ratios of $[M]^+ : [M+16]^+ = 100 : 0.3$ should occur in the corresponding hypothetical mass spectrum. The experimental results though do show a strong signal at $[M+16]^+$ and thus reaction of $O(^3P)$ directly with primary radical cations cannot account for the experimentally observed ion population as shown in Table 6.

b) Impact of H_2O, OH, and H on the ion distribution

As previously mentioned, no reactivity of the pyrene radical cation towards closed shell molecules such as H_2O was observed [70]. This is in very good agreement with the present experimental results. However, photodissociation of H_2O with the 193 nm laser pulse results in H-atom and OH radical mixing ratios of 0.6 ppbV, respectively. Unfortunately, no rate constant for the addition reaction of OH to $[M]^+$ is available. Atkinson et al. reported the rate constant for the analogue neutral reaction

$$\text{Pyrene} + \text{OH} \rightarrow [\text{Pyrene+OH}] \qquad (\text{rxn 22})$$

to be $k = 5 \cdot 10^{-11}$ $cm^3 \cdot molecule^{-1} \cdot s^{-1}$ [153]. Frequently, the rate constants for the same type of reaction can rise by one or two orders of magnitude if one of the reactants is charged. For example, the $O(^3P)$ addition to naphthalene is approximately 90 times faster when the aromatic system carries a positive charge [154,155].

If a rate constant of $k = 1.9 \cdot 10^{-9}$ $cm^3 \cdot molecule^{-1} \cdot s^{-1}$ for the OH addition reaction to the pyrene radical cation is assumed, a signal distribution of $[M]^+ : [M+OH]^+ = 100 : 95$ is calculated for a reaction time of 23 ms. Hence, a hypothetical mass spectrum would exhibit two signals, $m/z = 202$ and $m/z = 219$ with comparable intensities. This is in very good agreement with the experimental observation with respect to the signal intensity of the sum of the oxygenated products, i.e., $[M+15]^+$, $[M+16]^+$, and $[M+17]^+$, relative to $[M]^+$ (cf. Table 6). The already discussed consecutive H-atom abstractions leading to the final product distribution are driven by reactions with O_2, O_3, H and OH. All of these species react with H-atoms exergonic enough to compensate most of the positive Gibb´s free energies of dissociation for the C-H and O-H bond cleavages (cf. Table 5, No 12/14/15/16 compared to No 3/4/5/6).

4 Results and Discussion

In conclusion, the OH addition to the pyrene radical cation is determined to be the first step with a rate constant estimate of $k^7 = (1.9\pm0.9)\cdot10^{-9}$ cm^3·molecule^{-1}·s^{-1}, whereas the subsequent H-atom abstractions are straightforward, since all of the mentioned reactants are present in large quantities. This reasoning is supported by experimental data: Upon increasing any radical concentration, the relative distribution among the oxygenated products always led to dominating signals at $m/z = 217$, corresponding with formation of the most stable phenoxy type product (structures #8 - #10 in Figure 42).

The remaining species that has to be considered in this context is the H-atom. Le Page et al. [70] determined the rate constant of the addition reaction $[M]^+ + H$ to form $[M+H]^+$ as $k = 1.4\cdot10^{-10}$ cm^3·molecule^{-1}·s^{-1}. Thermochemical data as well support this reaction with Gibbs free energies of ΔG_R = -1.71 eV, -2.33 eV and -1.88 eV for position 1, 2 and 3, respectively. Applying the above rate constant would result in a relative ratio of $[M]^+ : [M+H]^+ = 100 : 5$ with 23 ms reaction time. The additional consideration of the isotopic ratios leads to a hypothetical outcome of the mass spectrum with a ratio of m/z 202 : 203 = 100 : 22; the experimentally observed ratio was 100 : 17. The present results are thus ambiguous. However, the back reaction of $[M+H]^+$ with O$_3$, OH and also H is exergonic. It appears to be unlikely that under the prevailing conditions any significant population of $[M+H]^+$ was thus produced.

c) Impact of Cl, ClO, and ClOO on the ion distribution

In the kinetic investigations described so far, chlorinated reactants were absent. Upon introducing chlorinated species, the ion chemistry became increasingly complex. The product analysis was thus rendered even more difficult. However, the impact of the presence of chlorinated species on the relative ion distribution is noteworthy. Feeding Cl-atoms, e.g., via photolysis of CH$_n$Cl$_{4-n}$ (n=0 – 2), to the reaction system as described in Table 6 had the following effects:

(i) Virtually no direct Cl-adduct formation $[M+Cl]^+$ or $[M-H+Cl]^+$ was observed, though the addition reactions were calculated to be exergonic in the range of ΔG_R = -0.22 eV up to -0.96 eV. Photolysis of 170 ppmV CCl$_4$ resulted in about 17 ppbV Cl-atoms (single laser pulse at 193 nm, 10 ns, 1 mJ). One reasonable explanation for missing Cl-adducts is the fast competing reaction No 10 (Table 4), which rapidly traps initially formed Cl-

---

[7] The error accounts for ± 5 ms reaction time.

4 Results and Discussion

atoms as ClOO, which at 1 bar rapidly decomposes to recycle Cl-atoms. Thus the concentration of Cl becomes quasi-stationary and might be too low to react swiftly enough with primary radical cations.

(ii) A ClO-adduct with the characteristic chlorine isotope pattern at masses $m/z = 253/255$ was observed when O_2 and Cl were present in large quantities. CID of $m/z = 253$ led to mass $m/z = 218$, assigned to $[M+O]^+$. It follows that the chlorine atoms react to form the relatively stable ClO-radical (via reactions no10 and no12, Table 4), which then readily adds to the aromatic ring system.

(iii) The presence of Cl-atoms drastically changed the relative intensity distribution among the oxygenated products, favoring the most stable phenoxy type species ($m/z = 217$) as observed before. Obviously, the H-abstraction reactions (No 3/4/5/6, Table 5), following the initial oxygen addition step, are strongly supported by the exergonic formation of HCl (No 17, Table 5).

The signal intensity of the sum of the oxygenated products relative to the intensity of the pyrene radical cation was also drastically changed, up to the complete loss of $[M]^+$. It is worth mentioning that it is unlikely that chlorine atoms or any other of the Cl containing radical species abstract an H-atom from the carbon skeleton of $[M]^+$ other than in the O-atom activated positions (cf. Figure 42 and ΔG_R Table 5, No 17/18/19 compared to No7). Thus the previously excluded second reaction pathway via the pyrenyl cation $[M-H]^+$ is also not favored in the presence of chlorine species.

In summary, Cl-atoms do not add directly to $[M]^+$ nor do they abstract H-atoms at any appreciable rates. Nevertheless, their presence strongly promotes the oxygenation of $[M]^+$ to yield $[M+O-H]^+$ via H-abstraction after the primary oxygenation step. Note that the absorption cross sections of chlorinated methanes are generally two orders of magnitude higher than of H_2O and more than four orders of magnitude higher than of O_2.

4.3.2.4 Consequences for Degradation Studies with APPI-MS

It is stressed that these results cannot be completely generalized. A summary of the most relevant neutral radical reactions that have to be considered when mechanistic studies are carried out in an API source are given in Table 4. However, this chemistry becomes increasingly complex when further compounds with high photodissociation rates in the VUV

4 Results and Discussion

are added in large quantities. Many radical species potentially affect the fate of primarily generated analyte ions and may open new ways of ion transformation processes. These processes become clearly visible at low primary ion concentrations (typical in the pptV range when the neutral analyte is present in the ppbV range) and might be much less observable when charged species are in excess. In this case, neutral radicals are merely titrated away and the remaining ion population is detected without notable transformation products present. Furthermore, the investigated reactions for pyrene apparently cannot be one-to-one adapted to other analytes since each compound has individual reaction mechanisms. For example, taking ozone not as the primary attacking species will probably not hold true for other analytes due to its large oxidizing potential and its major abundance (in the ppbV range when oxygen is present in the %-regime)[8] in typical APPI applications. Gas phase ozonolysis of double bonds is just one of the keywords to mention. Certainly, it is not of analytical interest to investigate individual reaction pathways caused by the analytical method itself and this was not the point of the previous section. However, it demonstrated the complexity of possible reactions, following the ionization step, which are typically not considered in APPI-MS applications. Furthermore, very good estimates can be derived concerning the analysis of samples from atmospheric degradation product studies with the novel APPI approach. In this way the constraints to which extent NRITPs can be reduced when applying the 0.13 ms ionization position were determinable. Figure 44 depicts a kinetic simulation for the reaction of a primary generated ion with a neutral radical species (denoted as *[NR]*) according to

$$[M]^+ + [NR] \rightarrow [P]^+ \quad\quad\quad (\text{rxn 23})$$

The rate constant was assumed to be $k = 2.0 \cdot 10^{-9}$ cm$^3 \cdot$molecule$^{-1} \cdot$s^{-1}, similar to the one obtained in the previous section. It is noted that this value is among the highest observed rate constants for ion molecule reactions [69]. The initial concentration of *[M]$^+$* and *[NR]* were 10 pptV and 10 ppbV, respectively, which are realistic values expected for degradation product study conditions. After 5 ms of reaction time nearly 90 % of the initial radical cation population is converted into a product species, whereas the 0.13 ms position is capable of preserving 94 % of the primarily generated *[M]$^+$*. In contrast to common APPI applications (minimum of 5 ms transfer time from ionization into the collision free region; cf. *4.2.2.3 LFIS – APPI, b*) the approach of ionizing on the transfer capillary virtually excludes interfering mass signals that are attributable to VUV-radiation initiated neutral radical ion

---

[8] Note that ozone is present by a factor of 1000 in excess relative to the primary ion concentration.

4 Results and Discussion

chemistry. This result was very well experimentally supported since NRITP based mass signals have never been observed with the novel APPI approach, even within the presence of air as the bulk gas.

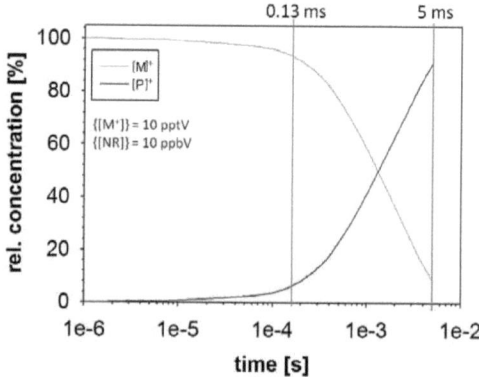

Figure 44: Kinetic simulation of a hypothetical reaction with $[M]^+ + [NR] \rightarrow [P]^+$, with a rate constant of $k = 2.0 \cdot 10^{-9}$ $cm^3 \cdot molecule^{-1} \cdot s^{-1}$ and with initial mixing ratios of $\{[M]^+\}$ = 10pptV and $\{[NR]\}$ = 10ppbV.

4.3.3 ITP via Chemical Ionization

The following chapter is concerned with ITPs in which charge of a primarily generated ion is transferred to a second species. Consequently, positive and negative charge transfer as well as protonation, deprotonation and ion clustering reactions are encompassed by the term "chemical ionization". At this point a short note on the general terminology used for secondary ionization methods is given. In classical low pressure mass spectrometry the term "chemical ionization" (cf. *1.2.1 Atmospheric Pressure Ionization (API)*) was used to describe the ion forming mechanism - ionization of analyte via a *chemical* reaction – but has been expanded to also express the prevailing primary ionization conditions. With the new ion sources operating at up to one bar pressure this term was extended to describe API methods, e.g., "atmospheric pressure chemical ionization (APCI)", but, again, the primary ionization conditions (corona discharge and atmospheric pressure) are directly connected with the method. However, independent of the primary ionization source and the prevailing pressure at the ionization step, the subsequent charging of an analyte is always a chemical ionization process. Consequently, in the case of ionization processes following the primary ionization

4 Results and Discussion

step, the terminology of the overall procedure should include the notation of the primary charge producing method and separately the notation describing the charge transfer. In that way, instead of using "APCI", the term should more likely be "atmospheric pressure corona discharge – positive chemical ionization (APCD-PICI)", with the prevailing pressure and the primary charge generation method denoted in the first term and the subsequent chemical ionization nature, including the type of charge being transferred, in the second. Along this line the prefix "dopant assisted" (cf. *1.2.1.2 Atmospheric Pressure Photo Ionization (APPI)*) is misleading since it generally describes a process in which primarily ionized species are added in *abundance* relative to the *small* amount of the neutral target analyte present in the sample flow, comparable to the addition of a reagent gas in classical CI. However, per definition the term "dopant" stands for the opposite case, in which a *small* amount of one species is added to a *large* amount of another to cause an effect. Again, the secondary process is merely a chemical ionization procedure, and thus in order to avoid any confusion and to correctly reflect the actual mechanism it should also be termed "chemical ionization". Accordingly, the overall process of, e.g. dopant assisted APPI, should then be denoted as "atmospheric pressure photo ionization – positive ion chemical ionization (APPI-PICI)" and "atmospheric pressure photo ionization – negative ion chemical ionization (APPI-NICI)" for the positive and negative mode, respectively. In the following the suggested terminology will be adapted

4.3.3.1 APPI/APLI-Positive Ion Chemical Ionization (PICI)

The protonation of analyte molecules through primary generated species (here via APPI or APLI) is the most prominent reaction belonging to this category of transformation processes. The example shown in section *4.2.3.4 Impact of Different Ionization Positions on MS Spectra, (b)* demonstrated the capability of ionized p-xylene to serve as the actively protonating agent for its own degradation product (cf. reaction 18, page 90). Upon ionizing on the 5 ms APPI position the observed spectrum clearly reflected the thermodynamically equilibrated ion distribution (cf. Figure 39, c). A similar picture was obtained upon analyzing the same degradation experiment with APLI. With respect to the 100 ms transit time within the laminar-flow ion source (average velocity of $v_x = 1.9$ m·s^{-1} and tube length of $l = 0.2$ m) the thermodynamic response within the APLI configuration was as expected. Obviously this type of transformation process clearly enhances the qualitative detection of degradation products that are otherwise not amenable to APLI. On the other hand the observed relative ion

4 Results and Discussion

distribution mostly does *not* reflect the neutral composition of the sample anymore. Consequently, also ionization within the present APLI configuration exhibited a much faster progression of the degradation experiment than it really was. It follows that kinetic studies of degradation experiments, in which ion chemistry is "mixed" into time dependent evolution of the main neutral species, are rendered nearly impossible.

Another investigated example for APPI-PICI is the well known self protonation reaction of acetone according to

$$[C_3H_6O]^+ + C_3H_6O \rightarrow [C_3H_6OH]^+ + C_3H_5O \qquad \text{(rxn 24)}.$$

A 1 ppmV acetone sample in pure nitrogen was delivered from the photoreactor. APPI was performed on the 5 ms position with the home-built discharge lamp (cf. Figure 32, left), separated from the sample gas flow with the LiF window (cf. Figure 24), and on the 0.13 ms position with the Kr-RF lamp and the LiF window mounted on the transfer capillary as shown in Figure 29, respectively. As depicted in Figure 45, (right), upon ionizing on the 5 ms position solely the quasi-molecular *[M+H]$^+$* ion at *m/z* 59 is present in the obtained mass spectrum.

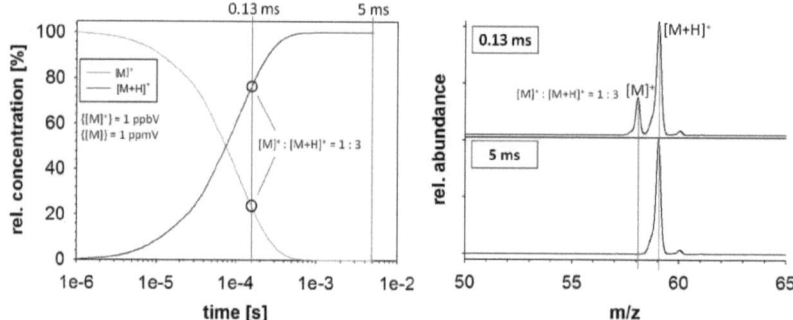

Figure 45: (left) Kinetic simulation of the self-protonation reaction of acetone according to [M]$^+$ + [M] → [M+H]$^+$ + [M-H], with a rate constant of $k = 3.8 \cdot 10^{-10}$ cm^3·molecule^{-1}·s^{-1} [156] and initial mixing ratios of 1 ppbV and 1 ppmV for the radical cation and the neutral acetone, respectively. (right) Recorded mass spectra upon ionizing 1 ppmV acetone in pure nitrogen on the 0.13 ms (top) and 5 ms (bottom) position, respectively.

In the mass spectrum recorded upon ionizing at the 0.13 ms position, however, the radical cation *[M]$^+$* at *m/z* 58 appeared with a relative abundance of *[M]$^+$* : *[M+H]$^+$* = 1 : 3. From a kinetical point of view this case is similar to the NRITP conditions in Figure 44, except that

4 Results and Discussion

the excess species (neutral acetone) is present in much higher concentration. The rate constant for this reaction was taken from MacNeil et al. [156] with $k = 3.8 \cdot 10^{-10}$ cm$^3 \cdot$molecule$^{-1} \cdot$s^{-1}. The initial mixing ratio for neutral acetone was very precisely determined with 1 ppmV. The initial mixing ratio of the radical cation $[M]^+$ was roughly estimated to be 1 ppbV. As explained in the preceding section, variations in the absolute concentration of $[M]^+$ do not change the relative distribution between $[M]^+$ and $[M+H]^+$ as long as the concentration of the neutral analyte is present in large excess. It follows that the entire reaction proceeds in pseudo first order with respect to the radical cation. Furthermore the well known dimerization and acetylation reactions of the quasi-molecular ion and the radical cation of acetone, leading to m/z 117 and m/z 101, respectively, were not observed at the prevailing conditions [156]. The result of the kinetic simulation is depicted by Figure 45, (left). Within 5 ms the entire radical cation population $[M]^+$ is transformed into the quasi-molecular species $[M+H]^+$, which is in good accord with the experimental data. Furthermore the simulation shows that after 0.13 ms the relative ion population of both species should be $[M]^+ : [M+H]^+ = 1 : 3$. Again this was very well supported by the experimentally obtained relative signal intensities as shown in Figure 45, (right, top). These results have several crucial consequences: (i) Chemical ionization processes at atmospheric pressure are fast and proceed quantitatively in common APPI configurations when the neutral species is present in large excess relative to the primary radical cation population, (ii) the approach of photoionizing on the transfer capillary is capable of preserving substantial amounts of radical cations that are potentially affected by chemical ionization ITPs (cf. *4.2.3.4 Impact of Different Ionization Positions on MS Spectra, b)*, however, (iii) the novel APPI method is on the limit of efficiently reducing ITPs based on chemical ionization, and (iv) the hypothesis of calculating dwell times within transfer capillaries using fluid dynamical expressions was once more substantiated due to the remarkable consistency between simulation and experiment.

4.3.3.2 APPI/APLI-Negative Ion Chemical Ionization (NICI)

The negative mode of APLI and APPI-MS is driven by negative chemical ionization ITPs. This holds true in particular for samples from atmospheric degradation experiments with 20 % oxygen present. Via three body collisions the photoelectrons from ionized analyte molecules are readily titrated away by O_2 to form the primary negative charged species O_2^- according to

4 Results and Discussion

$$O_2 + e^- + N_2/O_2 \rightarrow O_2^- + N_2/O_2 \qquad \text{(rxn 25)}.$$

The rate constant for this reaction was determined by Shimamori et al. to be in the range of $1.5 - 2.0 \cdot 10^{-30}$ cm^6·molecule^{-2}·s^{-1} [157]. A kinetic simulation with 1 ppbV as the initial mixing ratio of thermal electrons, 20.5 % O_2 and $k = 1.5 \cdot 10^{-30}$ cm^6·molecule^{-2}·s^{-1} revealed that within 0.2 µs all electrons are quantitatively attached to oxygen[9]. It follows that any obtained ion signal distribution in the negative mode emanates from the primary generated superoxide anion population. The electron affinity of O_2 is $E_a = 0.45$ eV and the gas phase basicity of O_2^- is $\Delta G_b = 15.0$ eV [138]. This principally opens up two alternative reaction pathways: (i) Charge transfer to species with higher electron affinities

$$O_2^- + [M] \rightarrow [M]^- + O_2 \qquad \text{(rxn 26)}$$

and (ii) deprotonation of neutral analyte molecules with lower gas phase basicity

$$O_2^- + [M] \rightarrow [M-H]^- + HO_2 \qquad \text{(rxn 27)}.$$

Both reactions have been observed for samples of degradation product studies. Reactions according to reaction 26 are of paramount importance for analytes containing nitro groups, as e.g. the electron affinity of 3,5-dimethylnitrobenzene was found to be $E_a = 1.21$ eV [158]. Thus degradation products of this type are in principal amenable to APPI-NICI. However, typical degradation experiments are performed with around 1-5 ppmV NO present (cf. *3.3.1 Procedure of Atmospheric Degradation Studies*) and additionally a fairly high concentration of NO_2 is build up during the experiment. Moreover, generated OH radicals further oxidize parts of NO_2 and NO to form nitric acid (HNO_3), as illustrated by reaction 14 (see page 88), and to nitrous acid (HNO_2) according to

$$NO + OH + N_2/O_2 \rightarrow HNO_2 + N_2/O_2 \qquad \text{(rxn 28)}$$

with a rate constant of $k = 1.5 \cdot 10^{-30}$ cm^6·molecule^{-2}·s^{-1} [138]. Consequently, fairly high background mixing ratios of HNO_3, HNO_2 and NO_2 are always present (ppmV range) in a sample from a degradation experiment. The electron affinity of NO_2 is $E_a = 2.3$ eV and the gas phase basicities of NO_2^- and NO_3^- are $\Delta G_b = 14.5$ and 13.7 eV, respectively [138]. In other words, these three species significantly impact the negative ion chemistry and the recorded ion signal distribution. The dominating reactions following the primary generation of the superoxide anion are thus expected to be in analogy to reactions 26 and 27, with

---

[9] Note that for a complete picture also the loss due to diffusion to the walls should be considered.

4 Results and Discussion

$$O_2^- + NO_2 \rightarrow NO_2^- + O_2 \quad \text{(rxn 29)}$$

and

$$O_2^- + HNO_3 \rightarrow NO_3^- + HO_2 \quad \text{(rxn 30)}$$

$$O_2^- + HNO_2 \rightarrow NO_2^- + HO_2 \quad \text{(rxn 31)},$$

respectively. According to the higher gas phase basicity of NO_2^- relative to NO_3^-, protonation subsequent to reaction 29 will occur

$$NO_2^- + HNO_3 \rightarrow NO_3^- + HNO_2 \quad \text{(rxn 32)}.$$

Moreover, most analyte species *[M]⁻* that have been charged shortly after the formation of the superoxide anion (reactions 26 and 27) will eventually be affected by charge annihilation according to reactions 15 - 17 (see page 88) or via charge transfer according to

$$[M]^- + NO_2 \rightarrow NO_2^- + [M] \quad \text{(rxn 33)}$$

of which the NO_2^- will subsequently be protonated according to reaction 32. As a result deprotonated nitric acid is speculated to dominate within the negative ion chemistry of such a sample. This speculation is very well supported by experimental data. With sufficient reaction time after the ionization step allowing the system to thermodynamically equilibrate (APLI or APPI on the 5 ms position) virtually no other signals than the *m/z* -62 and *m/z* -125, assigned to the NO_3^- and the cluster $[HNO_3 \cdot NO_3]^-$, respectively, are present in the recorded mass spectrum (cf. Figure 39, b, 5 ms). In contrast, with a kinetically controlled ion distribution, as demonstrated in cf. Figure 39 (b, 0.13 ms and 0.02 ms), reactions 15 - 17 (see page 88), 32 and 33 have not gone to completion. Equation 32 predicts an increase in the NO_2^- signal (*m/z* -46) with decreasing reaction time for the initiated negative ion chemistry. This is again very well supported by experimental data obtained when ionizing at different positions (cf. Figure 39, b).

4 Results and Discussion

4.4 Degradation Studies

4.4.1 Features and limitations of the MS setup

So far a modified mass spectrometric system for *in situ* monitoring of atmospheric degradation studies was setup and characterized. Features, advantages, as well as limitations of the entire assembly, i.e., the sampling unit, APLI in laminar flow ion source, APPI on the transfer capillary and the ion trap analyzer are briefly summarized as follows:

(i) The continuous gas flow from the photoreactor to the entrance of the transfer capillary of the MS is essentially laminar. Perturbation of the initial neutral sample composition via enhanced wall contact through turbulences is thus minimized. The time between sampling and analyzing accounts for ~0.5 s.

(ii) Selective and sensitive detection of aromatic hydrocarbons in the pptV range with APLI.

(iii) Nonselective and sensitive detection of diverse analytes in the lower ppbV range with windowless miniature spark discharge lamps mounted on the transfer capillary.

(iv) In negative ion mode it is possible to detect nitro compounds with APPI-NICI.

(v) Preservation of MS information due to kinetically controlled reduced ion transformation processes.

(vi) MS analysis of sample composition at different stages of the progressing ion chemistry (different ionization positions). This allows gathering of valuable information on ITPs from which also valuable information on the structural and chemical nature of compounds may be derived.

(vii) Quasi-simultaneous recording in the positive and the negative ion mode with a time resolution in the ms range.

(viii) Performance of carefully adjusted collision induced dissociation experiments.

(ix) The mass resolution is rather low, i.e., only nominal masses are used for reliable data interpretation.

(x) Due to the nature of the trap only a limited mass range is sensitively detected.

(xi) Exact quantitative data interpretation is rather limited, however, through the reduction of superimposed ion chemistry rough estimates of the relative composition of the neutral sample is possible.

4 Results and Discussion

4.4.2 Exemplary Degradation Study

4.4.2.1 Blank test without p-xylene present in the reactor

In the following a blank test was performed, with all chemicals (2.4 ppmV NO and 3.7 ppmV methyl nitrite) except p-xylene being injected into the photoreactor and backfill to 1000 mbar at 27 °C. The concentrations were determined by FT-IR analysis. Switching on all 32 superactinic fluorescent lamps started the chemical cascade initiated by the photodissociation of methyl nitrite according to reactions 7 – 9 (see page 28). A windowless miniature spark discharge lamp on the 0.13 ms position was used as the photoionizing source. Due to a prior experiment acetone was still present as an impurity (~3 ppbV) with the quasi-molecular ion as the most abundant peak at the beginning of the experiment and with a relative abundance[10] of 88 % at the end. The parameters of the MS were optimized for m/z 59 and applied for the positive as well as for the negative mode. Experience has shown that the resulting mass range sensitivity with these settings is between m/z ~30 – 250. Beyond these limits significant trapping and analyzing deficits have been observed.

a) Signals of protonated water clusters

Three signals in the positive mode, m/z 37, 55 and 73, were assigned to protonated water clusters of the type $[(H_2O)_nH]^+$ with n = 2 - 4. The intensities did not change during the entire blank test and the summed relative abundance accounted for approximately 12 %. The appearance of water clusters in this case is speculated to be due to the protonation reaction via the acetone radical cation:

$$(H_2O)_n + [C_3H_6O]^+ \rightarrow [(H_2O)_nH]^+ + C_3H_5O \qquad \text{(rxn 34)}$$

b) Signals of NO_x, HNO_x and $HNO_x \cdot NO_x$

As discussed above the presence of NO entails a rich NO_x and HNO_x ion chemistry which is clearly visible in the recorded mass spectra. Prior to initiating the photochemistry, nitrogen oxide appeared in the positive ion mode as NO^+ at m/z 30 with a relative abundance of 18 %. It did not appear in the negative mode due to the low electron affinity of

---

[10] In the following the signal abundances [%] represent the signal strength of a m/z value relative to the most abundant m/z signal of the mass spectrum.

4 Results and Discussion

$E_a = 0.03$ eV [138]. With the start of the photodissociation process of CH_3ONO, OH radicals partially oxidized NO to HNO_2 as shown in reaction 28 (see page 112). Consequently, the NO^+ signal decreased by 50 % during the blank test, as depicted in Figure 46.

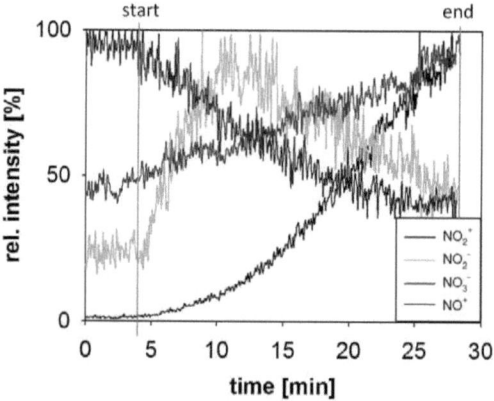

Figure 46: Blank degradation experiment with 2.4 ppmV NO and 3.7 ppmV methyl nitrite injected into the photoreactor and backfill to 1000 mbar with synthetic air. Shown are the obtained extracted ion chromatogram traces of NO^+, NO_2^+, NO_2^- and NO_3^-. "Start" and "end" denote the time where the superactinic fluorescent lamps were switched on and off, respectively.

Nitrogen dioxide was detected at m/z ±46 in the positive (4 % rel. abundance), as well as in the negative mode (2.8 % rel. abundance) as NO_2^+ and NO_2^-, respectively. Its presence, prior to the start of the blank test, was partially due to impurity within the compressed NO gas cylinder. Parts of the NO_2^- signal in the negative mode may also be explained by the presence of HNO_2 as an impurity from prior degradation experiments (cf. reaction 28, page 112), however, not detectable with the FT-IR anymore. Remarkable are the significantly differing slopes of the nitrogen dioxide signal in the positive and negative mode (cf. Figure 46). The NO_2^+ signal showed a steep increase during the degradation, as was expected according to reaction 9 (see page 28). FT-IR data supported this result as well, and a mixing ratio evolution from "non-detectable" up to 3 ppmV was determined. In the negative mode, however, the maximum signal of m/z -46 appeared after 6 minutes of degradation (7.8 % rel. abundance) and subsequently showed a nearly linear decrease. This effect is readily explained by parallel protonation (reaction 32, page 113) of NO_2^- with increasing HNO_3 concentration (cf. reaction 14, page 88). As explained before the latter species lead to the most abundant signal in the negative mode (m/z -62) and showed a nearly linear increase during the degradation.

4 Results and Discussion

The fairly high background concentration of nitric acid was assigned to an impurity from earlier experiments. In this context a signal appearing in the negative mode at m/z -93 is noteworthy. Its maximum relative abundance accounted for 0.2 % and strictly followed the shape of the NO_2^- signal. CID revealed only one fragment at m/z -46. This signal was assigned to the cluster $[HNO_2 \cdot NO_2]^-$ generated according to the ion molecule reaction

$$NO_2^- + HNO_2 \rightarrow [HNO_2 \cdot NO_2]^- \qquad \text{(rxn 35)}.$$

It follows that with decreasing NO_2^- population, caused by protonation with nitric acid (reaction 32, page 113), also the concentration of this ion cluster decreased. A second cluster formation was observed in the positive mode at m/z 76 with a maximum relative abundance of 6 %. It was assigned to $[NO_2 \cdot NO]^+$, generated according to

$$NO^+ + NO_2 \rightarrow [NO_2 \cdot NO]^+ \qquad \text{(rxn 36)}$$

or

$$NO + NO_2^+ \rightarrow [NO_2 \cdot NO]^+ \qquad \text{(rxn 37)}.$$

The extracted ion chromatogram exactly matched the shape of the NO^+ signal, which very well supported the above suggestion.

c) Signals of O_x

Prior to the initiated photochemistry, the O_2^- signal (m/z -32) was present with a relative abundance of 0.4 % and completely disappeared with the start of the degradation, quenched via reactions 29 – 31 (see page 113). Furthermore, O_3^- at m/z -48 was visible with a relative abundance of 1.4 % prior to the start of the degradation. However, the presence of ozone was not expected in the dark. Consequently, its appearance was caused by the discharge lamp with VUV photodissociation of oxygen into $O(^3P)$ followed by the reaction with O_2 (cf. Table 4, No 1 and 6). Ionization occurred through charge transfer from primary generated O_2^- since the electron affinity of ozone exceeds (E_a = 2.1 eV [138]) the value of oxygen by 1.65 eV. During the degradation experiment the m/z -48 signal exponentially decreased to 0.1 % rel. abundance. However, a slight increase in O_3 with increasing NO_2 concentration was expected according to the photostationary state relation [1]

$$[O_3]_{ss} = \frac{j \cdot [NO_2]}{k \cdot [NO]} \qquad \text{(eq 36)}$$

4 Results and Discussion

with k as the rate constant for the NO oxidation through ozone and j the photolysis rate of NO_2 to yield NO and $O(^3P)$. Again, this observation was caused by the parallel negative ion chemistry (reaction 33, page 113), or via charge transfer form O_3^- to NO_2, which eventually quenches the generated ozone anion population.

d) Signals of CH_3ONO and its degradation products

Methyl nitrite signals appeared in the negative mode as $[H_2CONO]^-$ at m/z -60 with a relative abundance of 89 % and exponentially dropped to 6.5 % during the experiment. Due to the photoinitiated degradation via reactions 7 – 9 (see page 28) the formation of formaldehyde was expected. Accordingly, two signals in the recorded mass spectra were observed in the positive mode at m/z 31 and m/z 61, assigned to the protonated monomer $[H_2COH]^+$ and the protonated dimer $[H_2CO\text{-}H\text{-}OCH_2]^+$ of formaldehyde. The extracted ion chromatograms showed nearly linear increase with relative abundances of 5 and 67 %, respectively, at the end of the degradation. FT-IR data revealed that these mass signals represented a formaldehyde concentration of 2.8 ppmV.

Table 7: Summary of observed mass signals during a blank degradation test with 2.4 ppmV NO and 3.7 ppmV CH_3ONO present in 1000 mbar synthetic air.

m/z	assigned compound	max. rel. abundance [%]
+30	NO^+	18.0
+31	$[H_2COH]^+$	5.3
-32	O_2^-	0.4
+37	$[(H_2O)H]^+$	7.0
+46	NO_2^+	100
-46	NO_2^-	7.8
-48	O_3^-	1.4
+55	$[(H_2O)_2H]^+$	2.2
-60	$[H_2CONO]^-$	88.8
+61	$[H_2CO\text{-}H\text{-}OCH_2]^+$	66.8
-62	NO_3^-	100
+73	$[(H_2O)_3H]^+$	3.3
+76	$[NO_2 \cdot NO]^+$	5.8
-93	$[HNO_2 \cdot NO_2]^-$	0.2

4 Results and Discussion

4.4.2.2 Degradation Study with p-xylene

The following section is focused on the comparison of the most abundant observed degradation products of p-xylene in the literature [5] and their determination with the present setup. The APPI assembly used was a windowless miniature spark discharge lamp at the 0.13 ms position on the transfer capillary. As chemicals 2.4 ppmV NO, 3.7 ppmV methyl nitrite and 0.9 ppmV p-xylene were injected into the photoreactor and then backfilled with synthetic air to 1001 mbar at 26 °C. The MS parameter settings were identical to the blank test run. At the end of the experiment (after 24 minutes) 70 % of the initial p-xylene was degraded.

As explained in the introduction, two primary OH radical initiated reaction steps of the atmospheric oxidation of aromatic hydrocarbons are proposed: (i) H-atom abstraction from present methyl groups with subsequent oxidation to the corresponding aldehydes and further to the carboxylic species, and (ii) OH-addition to the ring system with subsequent formation of the phenolic species or the initiation of a ring-opening cascade. The oxidation pathways of the main observed products in the literature are sketched in Figure 47. The compound names, the molecular weight, the observed m/z values and their recorded maximum relative abundances are given in Table 8.

Table 8: Summary of the main degradation products of p-xylene as observed in literature [5], with the assigned m/z values and the maximum relative abundances as recorded in this work.

structure #	compound name	molecular weight [Da]	observed m/z	max. rel. abundance [%]
1	p-xylene	106	+106	100
2	tolualdehyde	120	+119/121	7.7/11.9
3	p-toluic acid	136	-135	1.8
4	4-methylnitrobenzene	137	+137/-137	1.5/1.7
5	2,5-dimethylphenol	122	+122	7.8
6	2,5-dimethylnitrobenzene	151	-151	1.9
7	2,5-dimethyl-p-benzoquinone	136	-136	2.2
8	(E,Z)-3-hexene-2,5-dione	112	+112/113	1.9/12.2
9	2-methylbutenedial	98	+98/+99	0.5/2.7
10	2,5-dimethylfuran	96	+96/97	0.6/2.5
11	methylglyoxal	72	+73	2.5
12	glyoxal	58	not observed	not observed

4 Results and Discussion

The compound numbering is denoted with "#" and refers to the assigned structures in Table 8 throughout this chapter.

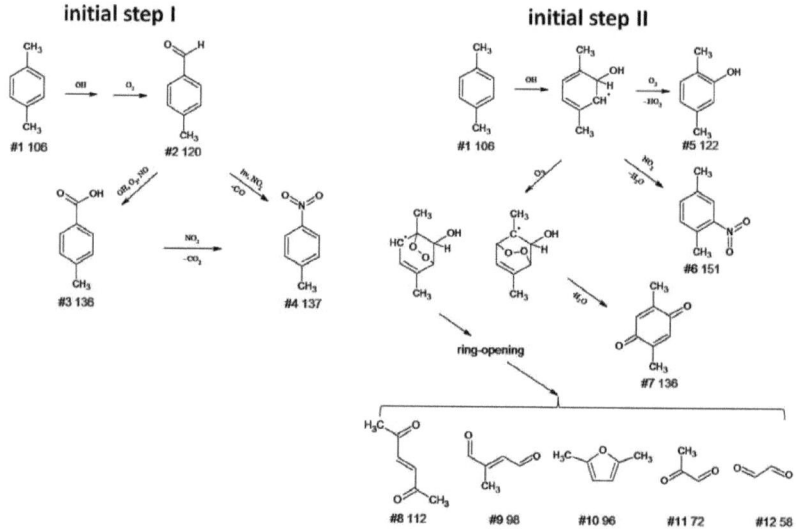

Figure 47: Schematic showing the oxidation pathways of the main observed degradation products of p-xylene as reported in the literature [5]. The red numbers denote the molecular weight in Dalton [Da].

a) Initial step I: H-atom abstraction from methyl group

As depicted by Figure 47 (left), tolualaldehyde is the observed primary product after the H-atom abstraction step. Its molecular weight is 120 Da, however, the most abundant signals assigned to this compound are m/z 119 ($[M+H]^+$) and 121 ($[M-H]^+$) with 7.7 % and 11.9 % relative abundance, respectively. This expected behavior is due to the instability of the radical cation. As confirmed through EI spectra [138], the unimolecular fragmentation with loss of an H-atom is the main process in a low collision environment. Furthermore CID on m/z 119 revealed the most abundant fragment ion at m/z 91, hence the loss of CO, which additionally supported its structural assignment. At atmospheric pressure, however, the $[M]^+$ may be subject to H-atom abstraction similar to the acetone radical cation (cf. 4.3.3.1 APPI/APLI-Positive Ion Chemical Ionization (PICI)). Furthermore, protonation due to the abundant presence of ionized p-xylene may have led to the appearance of the $[M+H]^+$. The

4 Results and Discussion

ionization potential of tolualaldehyde is $E_i = 9.33$ eV [138], thus not amenable to (1+1)-REMPI with the DPSS laser. This was very well confirmed by the absence of m/z 119 when the laser was used as radiation source. However, low abundance of m/z 121 was present due to the proposed protonation reaction through ionized p-xylene, whereas the appearance of m/z 119 requires the photoionization as the primary step. Tolualaldehyde is not observed in the negative mode, which is readily explained by its low electron affinity of $E_a = 0.37$ eV and high gas phase basicity $G_b = 15.3$ eV [138] of the deprotonated molecule. The evolution of the m/z 119 trace during the degradation experiment is depicted in Figure 48. Tolualaldehyde may further be oxidized to the p-toluic acid, as shown in Figure 47 (left). This product was observed in the negative mode as the $[M-H]^-$ at m/z -135 due to the relative low gas phase basicity of the deprotonated molecule ($G_b = 14.5$ eV). The extracted ion chromatogram is shown in Figure 48.

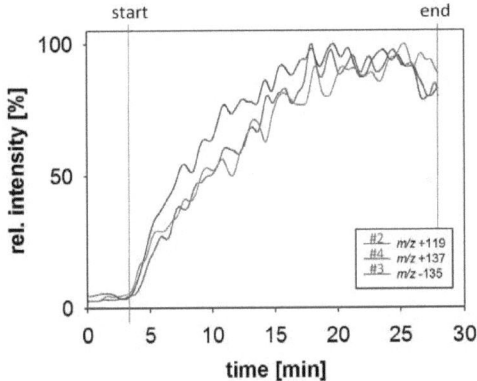

Figure 48: Degradation of p-xylene with 2.4 ppmV NO, 3.7 ppmV methyl nitrite and 0.9 ppmV p-xylene injected into the photoreactor and backfill to 1001 mbar with synthetic air. Shown are the obtained extracted ion chromatogram traces of tolualaldehyde (#2), p-toluic acid (#3) and 4-methylnitrobenzene (#4). "Start" and "end" denote the time where the superactinic fluorescent lamps were switched on and off, respectively.

It is noted that this value is identical to the gas phase basicity of NO_2^-. It follows that the shape of the chromatogram might have been partially affected by protonation through generated HNO_3 (see above). However, charge annihilation ITP is expected to be of minor importance due to the excess of NO_2^- (quenching effect of HNO_3) and thus the chromatogram

4 Results and Discussion

trace consequently is confirming that this species is a secondary product of the tolualaldehyde. The third observed compound belonging to this reaction pathway is the 4-methylnitrobenzene, formed by the species #2 and #3 via loss of CO and CO_2, respectively. It is observed in the presence of sufficient NO_2, with a minimum critical mixing ratio of around 300 ppbV, as determined by Seinfeld and Pandis [1]. FT-IR analysis revealed an increase of nitrogen dioxide of up to 3 ppmV in the present study. Methylnitrobenzene was observed in the negative as well as in the positive mode at $m/z \pm 137$ with 1.5 and 1.7 % relative abundance, respectively. The appearance in the negative mode is readily explained by the electron affinity of $E_a = 0.95$ eV [138], ensuring sufficient charge transfer from O_2^-. The shape of the extracted ion chromatogram again confirms #4 as a secondary product from tolualaldehyde (cf. Figure 48).

b) Initial step II: OH-addition to the aromatic ring

The primary generated OH-adduct may follow two subsequent reaction pathways: (i) H-atom abstraction leading to degeneration of the aromaticity and (ii) O_2 or NO_2 addition to the radical position of the ring system, provided that a sufficient concentration of nitrogen dioxide is present (see above). The first pathway eventually leads to 2,5-dimethylphenol, which was clearly observed in the positive mode at m/z 122 with a maximum relative abundance of 7.8 %. With laser ionization a relative abundance of 9.7 % was obtained, which very well confirms the assignment of this peak to a ring-retaining product. Furthermore, the ion chromatogram trace in Figure 49 (left) shows a steep increase shortly after the photoinitiation of the degradation cascade, reaches a maximum after seven minutes and then decreases as a result of subsequent oxidation steps. The shape of the curve demonstrates that structure #5 is one of the first generated species along the oxidation pathway. In contrast, the extracted ion chromatogram for m/z -151 (2,5-dimethylnitrobenzene) shows a less steep increase with a plateau reached in the end of the experiment (cf. Figure 49, left), accounting for 1.9 % relative abundance. This behavior is readily explained by the gradual increase of NO_2 during the degradation, as can be seen in the chromatogram trace of m/z 46 (NO_2) in Figure 49 (right). In this context a peak at m/z -152 with 9.6 % maximum relative abundance is noted. This signal is proposed to be the result of superimposed ion clustering chemistry according to

$$\text{p-xylene} + NO_2^- \rightarrow [\text{p-xylene} + NO_2]^- \qquad \text{(rxn 38)}.$$

4 Results and Discussion

This assumption was supported by the match of the cluster chromatogram trace with the steep, virtually linear curve progression of NO_2 (cf. Figure 49, right). In contrast hereto the occurrence of a plateau or an intermediate maximum is expected for a photoinititated degradation product as shown for 2,5-dimethylnitrobenzene, cf. #6 in Figure 49. In the case of oxygen addition to the free radical position on the ring skeleton, further intermediate states, such as O-O-bridged species are proposed (cf. Figure 47, right). One subsequently generated product is assumed to be 2,5-dimethyl-p-benzoquinone with a molecular weight of 136 Da. The electron affinity of this compound is $E_a = 1.76$ eV [138], which makes it amenable to the negative mode due to charge transfer. The shape of the extracted ion chromatogram at m/z -136 is characterized by a maximum after about nine minutes of degradation time and subsequently decreases, indicating further oxidation (cf. Figure 49, left).

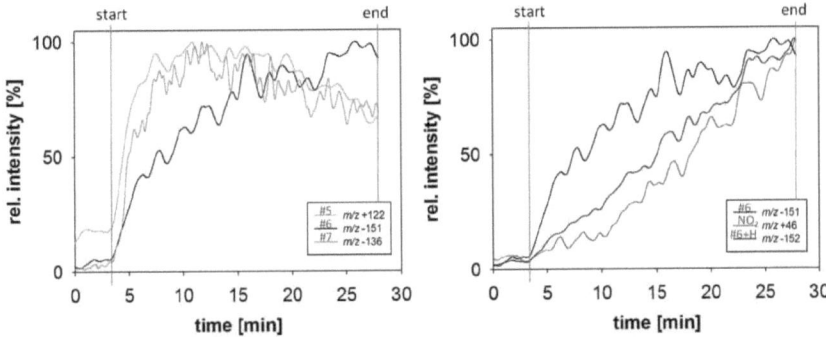

Figure 49: Degradation of p-xylene with 2.4 ppmV NO, 3.7 ppmV methyl nitrite and 0.9 ppmV p-xylene injected into the photoreactor and backfill to 1001 mbar with synthetic air. Shown are the extracted ion chromatogram traces of (left) 2,5-dimethylphenol (#5), 2,5-dimethylnitrobenzene (#6), 2,5-dimethyl-p-benzoquinone (#7) and (right) 2,5-dimethylnitrobenzene (#6), NO_2 and the [p-xylene·NO_2]⁻ cluster at m/z -152. "Start" and "end" denote the time where the superactinic fluorescent lamps were switched on and off, respectively.

The oxygen addition step to the free radical position moreover initiates a possible reaction cascade via ring-opening, as sketched in Figure 47 (right). Some of the main compounds described in the literature [5] (cf. Figure 47, #8 - #12) will be discussed briefly in the following. The dominating degradation product belonging to this class was 3-hexene-2,5-dione as E or Z isomer. As demonstrated in chapter *4.2.3.4 Impact of Different Ionization Positions on MS Spectra, b*, this compound was subject to significant protonation through

4 Results and Discussion

ionized p-xylene. Within this experiment the radical cation at m/z 112 and the quasi-molecular ion at m/z 113 were observed with maximum relative abundances of 1.9 and 12.2 %, respectively. The chromatogram trace of m/z 112, as depicted in Figure 50, is characterized by a steep increase with a maximum after around seven minutes degradation time, followed by a decline to finally reach 50 %. This curve shape is typical for compounds formed in an early stage of the oxidation process (compounds #5 and #7). On the contrary, the extracted ion chromatogram traces of structures #9 - #11 show typical behavior of secondary formed degradation products with a clear induction time, mostly ending in a plateau (cf. Figure 50). Structure #9 represents 2-methylbutenedial with a molecular weight of 98 Da. Observed were the radical cation (m/z 98) and the quasi-molecular ion (m/z 99) in the positive mode with maximum relative abundances of 0.5 and 2.7 %, respectively. In Figure 50 the recorded radical cation chromatogram trace is shown.

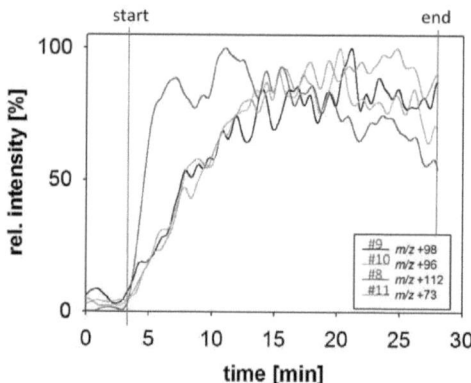

Figure 50: Degradation of p-xylene with 2.4 ppmV NO, 3.7 ppmV methyl nitrite and 0.9 ppmV p-xylene injected into the photoreactor and backfill to 1001 mbar with synthetic air. Shown are the obtained extracted ion chromatogram traces of (E,Z)-3-hexene-2,5-dione (#8), 2-methylbutenedial (#9), 2,5-dimethylfuran (#10) and methylglyoxal (#11)

Structure #10 represents 2,5-dimethylfuran, assigned in the present experiments to the signals m/z 96 ($[M]^+$) and m/z 97 ($[M+H]^+$), with relative abundances of 0.6 and 2.5 %, respectively. Figure 50 shows the radical cation chromatogram trace of this species. Methylglyoxal, structure #11, is solely observed as the quasi-molecular ion at $m/z = 73$ with a relative abundance of 2.5 %. CID revealed the loss of CO and CH_3 as the most important fragmentation process, which excludes the possible detection of the water cluster $[(H_2O)_3H]^+$ with the same nominal mass. #12 is the last structure shown in Figure 47 (right) representing the degradation compound glyoxal. In the literature [5] it is one of the most prominent observed species with up to 20 % of all degradation products. However, glyoxal was never observed in any of the present experiments, neither in the positive mode as radical cation (*m/z* 58) or quasi-molecular ion ($[M+H]^+$), nor in the negative mode at *m/z* -58 or as $[M-H]^-$ at *m/z* -57. This is surprising, since the ionization energy of $E_i = 10.2$ eV and the electron affinity of $E_a = 0.62$ eV [138] strongly suggest that glyoxal, if present, should be detected in both modes. So far, also no other signal that might represent a possible ion cluster of glyoxal was found. Furthermore, glyoxal was not observed in FT-IR data as well, however, this might be due to the generally less sensitive detection method than the present API-MS setup. Hence the potential fate of glyoxal within the present mass spectrometric analysis is still of question.

This chapter demonstrated the approach of data interpretation with this API-MS setup. The effects and parameters that need to be considered and that are used to obtain valuable structural as well as time dependent information regarding a degradation product were pointed out.

5 Summary and Conclusion

a) Investigations on the commercially available API source

Comprehensive studies of the fluid dynamical behavior of the commercially available atmospheric pressure ionization source revealed that mostly chaotic, turbulent flows prevail in the enclosure, causing long residence times up to seconds. This inevitably leads to uncontrolled hetero- and homogenous neutral- and ion transformation processes, to ion losses and possibly memory effects. Mechanistic studies are thus rendered nearly impossible. It was found that the conventional assembly additionally restricts the ionization volume and leads to an insufficient use of the ionizing laser radiation in APLI applications. Moreover the performance of APPI in such a source enclosure is virtually restricted to secondary ionization mechanisms due to the confined range of VUV radiation.

b) Development of a laminar flow ion source with a laminar sampling unit

Consequently, a new atmospheric pressure ionization source and sampling unit that are characterized by controlled, laminar flows were designed. Herein the probe from the reactor is directed at a 10° angle into the main tube of 20 cm length and 4 mm inner diameter, ending in a conically shaped outlet port that is directly connected to the transfer capillary of the mass spectrometer. The sampling flow accounts for 1.4 $L \cdot min^{-1}$, determined by the choked flow of the capillary. For balancing the pressure of the reactor a continuous flow of synthetic air is provided, exactly matching the MS sampling flow. This results in dilution of merely 8 % after 60 minutes, which is negligible within a degradation study. Fluid dynamical simulations on this source design substantiated the assumed laminar characteristic and additionally gave valuable insight into the time dependent and diffusion based spatial evolution of a generated confined ion packet along the propagation direction. The results of the latter supported preliminary experiments in which an efficient, long-distance ion transport within a laminar gas stream was observed. This eventually led to the development of a new APLI setup where the laser beam is directed coaxially along the flow direction of the gas sample into the main tube. Consequently, a significant increase of the ionization volume compared to the common APLI setup is achieved. Moreover, the laser pulse repetition rate of 200 Hz allows for manifold irradiation of a neutral analyte molecule before entering the MS. This approach establishes an APLI-MS method that allows the use of a small UV solid state laser (DPSS) as

5 Summary and Conclusion

radiation source with a comparable performance obtained as with an exciplex laser in current APLI-MS applications. The benefit is obvious: A DPSS laser significantly outperforms an exciplex laser in terms of size, noise level, maintenance, and purchase cost. Preliminary experiments showed lower detection limits for aromatic hydrocarbons in the pptV range. The modular design of this API source additionally allowed for the implementation of a VUV radiation unit that, in the first instance, made use of a specially shaped LiF window to maintain the laminar flow. Since the inner tube diameter accounts for 4 mm, sufficient VUV radiation for primary ionization of the analyte is provided. The benefits of this new source design are summarized as: (i) Controlled flow, (ii) high ion transmission efficiency into the MS, (iii) efficient irradiation of the sample (APLI and APPI), (iv) the modular design allows for versatile add-ons and for easy cleaning, and (v) the rather simple computed fluid dynamics enables superimposed kinetic calculations, which will eventually lead to valuable comparisons of simulation and experiment in the near future. At present, only gas phase samples are analyzable, however, the application of low-flow liquid nebulization stages is under investigation. Moreover, first approaches of implementing other ionization techniques such as APCI gave promising results.

c) Novel APPI approach with home-built miniature spark discharge lamps

Experiments with the laminar flow ion source demonstrated that, even with the VUV radiation inlet unit at the closest possible ionization position, a thermodynamic response of the system in terms of ion transformation processes resulted in most cases. On the one hand this can be desirable, since it can significantly enhance the sensitivity towards some compounds. On the other hand, the more the ion distribution reaches its thermodynamic equilibrium the more likely this situation does not reflect the unknown composition and may lead to severe misinterpretation and loss of mass spectrometric information. Accordingly, a completely new approach for APPI was developed within this work. This approach is based on VUV irradiation of the analyte gas flow within the transfer capillary, which is functioning as a pressure reduction unit between the atmospheric pressure and the low pressure region of the mass spectrometer. In this way the time between ionization and entrance into the collision free region of the mass spectrometer was reduced by a factor of 250 (from 5 ms in the VUV inlet unit of the LFIS to 0.02 ms at the exit of the capillary in the first differential pumping stage), whereas the analyte density was merely diminished by a factor of four. However, this

5 Summary and Conclusion

approach required obtaining open access to the capillary gas flow. Consequently, a characterization with respect to the fluid dynamical behavior and ion transmission efficiencies was required. Along this line it was shown that well known empirical fluid dynamical equations, developed for large tube systems, can be applied to the small dimensions of a transfer capillary. For example, the calculated static pressures were confirmed to be within ± 2 % accuracy of the experimentally observed values. The first VUV radiation inlet design made use of a specially shaped LiF window, mounted on the transfer capillary. As radiation source a commercially available Kr-RF lamp was used. The expected behavior in terms of reduced ion transformation processes was observed, however, sensitivity was not satisfactory, due to the ineffective use of the rather large radiation spot. Based on this result windowless miniature DC spark discharge lamps, directly mounted onto the transfer capillary, have been developed to provide precisely positionable ionizing radiation on a small illuminated area. Several different designs were introduced and the development of an appropriate high voltage power supply, tailored to the discharge characteristics of the lamp designs, was described. The type of discharge applied was characterized as a high pressure (p_{argon} = 200 - 1000 mbar), low gas flow (0.1 - 0.5 L·min^{-1}), medium frequency (1.5 - 2.0 kHz), and medium voltage (500 – 1500 V) DC spark with pulse currents of 2 A and pulse durations of 7 µs. The generated plasma is in a non-thermal and a non-equilibrium state which consequently results in mostly spectral line emission. The used discharge gas was argon and comprehensive experimental as well as theoretical investigations on the prevailing discharge chemistry and generated VUV radiation was given. It was shown that a separation of the discharge gas from the sample gas stream in the capillary is readily accomplished by adjusting the discharge pressure to the local static pressure. Moreover it was demonstrated that this design is fairly unique for providing ionizing radiation at elevated pressures even below the cutoff transmission (105 nm) of the commonly used LiF windows. So far, lower detection limits of 0.5 ppbV for benzene and 0.1 ppbV for 2-butanone have been determined. An increase of the ionization efficiency by a factor of 400 as compared to the first approach with the commercially available APPI lamp mounted on the LiF window equipped capillary was achieved. However, this approach still leaves room for further investigations in terms of optimizing and adjusting the discharge and radiation characteristics with, e.g. gas mixtures, electrode shape, etc. It is expected that strongly decreased detection limits are realizable.

For the self protonation reaction of acetone and for negative ion mass spectra obtained from an atmospheric degradation product study it was furthermore demonstrated that this

5 Summary and Conclusion

novel APPI approach was capable of revealing a kinetically determined ion distribution resulting in valuable MS information on the neutral analyte distribution.

d) Ion transformation processes

A comprehensive discussion was given on the occurrence of ion transformation processes. It was pinpointed that ITPs in general may prove valuable mass spectrometric information. However, and this was stressed, in order to correctly interpret a mass spectrum with respect to the neutral analyte distribution, it is necessary to know what kind of transformation processes between the ionization step and the detection step occurred and also to which extend. In API-MS this issue generally complicates the mass spectrometric analysis of unknown samples. When no reference spectrum is available illustrating an ITP reduced ion signal distribution, hardly any reliable analysis is possible.

It was demonstrated that kinetic energy increase of the travelling ions due to the presence of electrical field gradients along the ion transfer path to and within the analyzer (ion trap) can significantly impact the recorded ion distribution in terms of unintended induced unimolecular fragmentation processes.

Furthermore a comprehensive discussion was given on neutral radical induced ion transformation processes. Starting with the pyrene radical cation an intense experimental and theoretical investigation on the oxidation pathway with neutral radicals, generated through VUV radiation within an API source, was performed. Herein the analogy to well known atmospheric oxidation processes and the complex neutral radical chemistry became apparent. This section was thus of paramount importance since the present analytical system is supposed to monitor the neutral oxidation product distribution from a degradation study and hence superimposed ion oxidation processes would lead to significant misinterpretation. However, through kinetic calculations and experimental observations the occurrence of VUV initiated neutral radical induced ion oxidation can be excluded using ionization positions along the transfer capillary. It is noted that NRITPs are not only limited to APPI. Moreover any type of API approach using discharges at atmospheric pressure and leaving sufficient time between the ionization and detection step are in *generally* affected. In this context significant oxidation products can be found in the literature mass spectra as, e.g. shown by Shelley et al. [159], who used DART (direct analysis in real time) for ionization of ferrocene and observed

5 Summary and Conclusion

the oxygenated species with up to 15 % relative abundance, or as shown by Syage et al. [160], who recorded oxgenated products of pyrene with up to 50 % relative abundance using APCI.

A principal note was given on the literature terminology for the recently introduced types of ITPs. Herein any so called "dopant assisted" method was reduced to the terms positive ion chemical ionization (PICI) and negative ion chemical ionization (NICI) to more likely reflect the operating fundamental secondary chemical ionization mechanisms. The primary ionization step was suggested to be added as a prefix, as, e.g. APPI-PICI. The most important positive chemical ionization process is the protonation reaction. Exemplary on the self protonation reaction of acetone it was shown that (i) chemical ion transformation processes at atmospheric pressure are generally fast and proceed quantitatively in common APPI configurations when the neutral species is present in large excess relative to the primary radical cation population, (ii) the approach of photoionizing on the transfer capillary is capable of preserving substantial amounts of radical cations that are principally affected by subsequent ITPs, and (iii) the novel APPI reduces ITPs to a maximum extent. For the negative MS mode it was shown that at typical degradation product study conditions with 20.5 % oxygen present, the thermal electron capture occurs nearly quantitatively within 0.2 µs to form O_2^-. Hence any type of NICI process, i.e., negative charge transfer or deprotonation consequently emanates from the primary generated superoxide anion population. A comprehensive discussion on the most important subsequent ion molecule reactions for samples from degradation studies, in particular with respect to NO_x, HNO_x and O_x, was given.

e) Exemplary degradation study of p-xylene

The final chapter described the exemplary application of the novel setup for a degradation study of p-xylene. Therefore well known oxidation products and their detection within the present experiment were discussed. It was shown that the MS methods APLI, the novel APPI approach, APPI/APLI-PICI, APPI/APLI-NICI and CID combined with FT-IR data provide a powerful tool for *in situ* monitoring of photoinitiated atmospheric degradation experiments. In this way a broad range of diverse compounds can be analyzed in parallel with their time dependent concentration profile recorded. Furthermore, the approach of data interpretation with this setup was demonstrated. The effects and parameters that have to be considered and that can be used to obtain valuable structural as well as time dependent information on the occurrence of a degradation product were pointed out.

6 Indexes

6.1 List of Figures

Figure 1: (left) Atmospheric "washing machine". (right) Excerpts of the photo oxidation pathway of p-xylene. 2

Figure 2: Fundamental processes of APLI, APPI, and NIF. 7

Figure 3: Sketch of the general challenge in mass spectrometry, of how a mass spectrum can truly reflect the neutral composition. 8

Figure 4: (left) Photograph of the esquire6000. (right) Scheme of the different vacuum stages in the esquire6000, including a pressure diagram. 12

Figure 5: (left) Schematic of the ion trap. (right) Mathieu stability diagram. 14

Figure 6: Photograph of the laser systems used. 19

Figure 7: (left) Schematic of the Apollo/ MPIS source [48]. (right) List of the gas flows through the source enclosure. 21

Figure 8: Schematic of the novel laminar-flow ion source. 22

Figure 9: Home-built APPI setup within the transfer capillary. 23

Figure 10: Setup for fluid dynamical and ion transmission characterization of transfer capillaries. 24

Figure 11: Home-built ionization setup for kinetic and mechanistic studies [82]. 27

Figure 12: Schematic of the experimental setup used for degradation studies. 29

Figure 13: DIA plot obtained with common ion source parameter settings for routine LC-APLI using the MPIS [44]. 32

Figure 14: Computational fluid dynamical calculation results (Ansys CFX-11) of (left) the velocity distribution, (center) time integrated trajectories and (right) the neutral analyte distribution within the geometries of a MPIS based on typical source settings [48]. 34

Figure 15: Standard addition method to determine the oxygen mixing ratio in the present API source. 36

Figure 16: Fluid dynamical simulation of a tube with a volume flow of 1.4 $L \cdot min^{-1}$, inner tube diameter of 9 mm and an 0.8 mm orifice [48]. 38

Figure 17: Preliminary test on ion transport efficiencies in a laminar flow [48]. 38

6 Indexes

Figure 18: Recorded ion current as function of the laser frequency and the number of segment illumination within a coaxial setup. 41

Figure 19: (left) Standard addition method for anthracene in a coaxial APLI configuration. (right) Coaxial APLI mass spectrum with 3 pptV anthracene present. 42

Figure 20: Fluid dynamical simulation of the LFIS. 44

Figure 21: Simulated spatial evolution of an initially localized ion packet along the downstream propagation of the LFIS. 46

Figure 22: Power plot of the ion signal versus the photon flux Φ of the DPSS laser. 50

Figure 23: Time resolved current measurements using the laminar flow ion source. 52

Figure 24: Schematic of the APPI unit of the LFIS. 53

Figure 25: Results of an ion trajectory calculation for the transfer efficiency from the capillary through the skimmer. 54

Figure 26: Flow dynamical characterization of a home-made and an original transfer capillary. 56

Figure 27: Static pressure and velocity distribution within the transfer capillary. 60

Figure 28: Calculated plots of (left) the choked flow and (right) the critical pressure as function of the upstream pressure. 61

Figure 29: Photographs of (left) the LiF window mounted on the transfer capillary and (right)"APPI on transfer capillary" with the commercially available Kr-RF low pressure discharge lamp. 62

Figure 30: (left) Circuit diagram of the DD20_10 C-Lader power supply. (right) Photograph of the power supply. 65

Figure 31: (left) Transmission curves of a 5 mm and a 1 mm thick LiF window. (right) Comparison of mass spectra recorded upon ionization with and without LiF window separation. 67

Figure 32: Schematics of home-built spark discharge lamps, (left) design 1, (center) design 2, (right) design 3. 70

Figure 33: Calibration curves for the determination of the lower limit of detection (LOD) with the novel APPI setup. 74

Figure 34: Spark discharge characteristics of lamp design 3 operated with the DD20_10 C-Lader. 75

Figure 35: Typical Paschen curve for argon. 77

Figure 36: Mean partial electron energy contributions to the processes of ionization of the discharge gas, electronic excitation of the discharge gas, and kinetic energy increase of the electron stream within a discharge in pure argon.................80

Figure 37: (left) VUV spectra of the spark discharge. (right) UV/VIS spectrum of the spark discharge...............82

Figure 38: VUV measurement of a spark discharge lamp below 105 nm (in cooperation with Resonance Ltd. Barrie, On, Canada)...............87

Figure 39: Mass spectra obtained at different ionization positions...............89

Figure 40: Impact of the "trap drive" parameter on induced fragmentation of ionized p-xylene...............92

Figure 41: Mass spectra recorded upon irradiation of a nitrogen carrier gas flow in the presence of ~ 6 ppbV pyrene, ~170 ppmV CCl_4, ~100 ppmV O_2, and ~ 200 ppmV H_2O...............95

Figure 42: Possible structures within the oxidation pathways of the pyrene radical cation....97

Figure 43: Gibbs free enthalpies for the oxidation pathway of the pyrene radical cation.......99

Figure 44: Kinetic simulation of a hypothetical reaction with $[M]^+ + [NR] \rightarrow [P]^+$...........108

Figure 45: Kinetic simulation of the self-protonation reaction of acetone...............110

Figure 46: Blank degradation experiment...............116

Figure 47: Schematic showing the oxidation pathways of the main observed degradation products of p-xylene as reported in the literature [5]...............120

Figure 48: Degradation products of p-xylene I...............121

Figure 49: Degradation products of p-xylene II...............123

Figure 50: Degradation products of p-xylene III...............124

6.2 List of Tables

Table 1: Parameters of the used laser radiation sources...............19

Table 2: Calculated dwell times for ions within the capillary as function of the ionization position...............60

Table 3: Observed Ar I emission lines between 200 and 1100 nm...............84

Table 4: Neutral radical reactions considered in the present investigation...............94

Table 5: Computed ΔG_R values [eV] for the examined reactions...............98

6 Indexes

Table 6: Major peak abundances in a mass spectrum recorded upon irradiating a flow of N_2 at 1 bar with 420 ppmV H_2O, 0.7 % O_2 and 6 ppbV pyrene present. 102

Table 7: Summary of observed mass signals during a blank degradation test with 2.4 ppmV NO and 3.7 ppmV CH_3ONO present in 1000 mbar synthetic air. 118

Table 8: Summary of the main degradation products of p-xylene. 119

6.3 References

(1) Seinfeld, J. H.; Pandis, S. N. *Atmospheric Chemistry and Physics - From Air Pollution to Climate Change*; John Wiley and Sons, Inc.: New York, NY, USA, **1998**.

(2) Atkinson, R.; Arey, J. Atmospheric Degradation of Volatile Organic Compounds *Chemical Reviews* **2003**, *103*, 4605-4638.

(3) Kurtenbach, R.; Brockmann, K. J.; Lörzer, J.; Niedojadlo, A.; Becker, K. H. *VOC-Measurements in urban air of the city of Wuppertal*, University of Wuppertal, **1998**.

(4) Smith, D. F.; McIver, C. D.; Kleindienst, T. E. Primary Product Distribution from the Reaction of Hydroxyl Radicals with Toluene at ppb NOx Mixing Ratios *Journal of Atmospheric Chemistry* **1998**, *30*, 209-228.

(5) Calvert, J.G.; Atkinson, R.; Becker, K.H.; Kamens, R.M.; Seinfeld, J.H.; Wallington,T.J.; Yarwoods, G. *The mechanisms of atmospheric oxidation of aromatic hydrocarbons*; Oxford University Press, USA, **2002**.

(6) Arey, J.; Obermeyer, G.; Aschmann, S. M.; Chattopadhyay, S.; Cusick, R. D.; Atkinson, R. Dicarbonyl Products of the OH Radical-Initiated Reaction of a Series of Aromatic Hydrocarbons *Environmental Science & Technology* **2008**, *43*, 683-689.

(7) Zhao, J.; Zhang, R.; Misawa, K.; Shibuya, K.Experimental product study of the OH-initiated oxidation of m-xylene *Journal of Photochemistry and Photobiology A: Chemistry* **2005**, *176*, 199-201.

(8) Yu, J.; Jeffries, H. E.; Sexton, K. G. Atmospheric photooxidation of alkylbenzenes—I. Carbonyl product analyses *Atmospheric Environment* **1997**, *31*, 2261-2280.

(9) Volkamer, R.; Klotz, B.; Barnes, I.; Imamura, T.; Wirtz, K.; Washida, N.; Becker, K. H.; Platt, U. OH-initiated oxidation of benzenePart I. Phenol formation under atmospheric conditions *Physical Chemistry Chemical Physics* **2002**, *4*, 1598-1610.

(10) Klotz, B.; Volkamer, R.; Hurley, M. D.; Andersen, M. P. S.; Nielsen, O. J.; Barnes, I.; Imamura, T.; Wirtz, K.; Becker, K. H.; Platt, U. OH-initiated oxidation of benzenePart II. Influence of elevated NO x concentrations *Physical Chemistry Chemical Physics* **2002**, *4*, 4399-4411.

(11) Pandis, S. N.; Harley, R. A.; Cass, G. R.; Seinfeld, J. H. Secondary organic aerosol formation and transport *Atmospheric Environment. Part A. General Topics* **1992**, *26*, 2269-2282.

(12) University of Leeds The Master Chemical Mechansim http://mcm.leeds.ac.uk/MCM/home.htt **2010**.

(13) Derwent, R. G.; Jenkin, M. E.; Saunders, S. M.; Pilling, M. J. Photochemical ozone creation potentials for organic compounds in northwest Europe calculated with a master chemical mechanism *Atmospheric Environment* **1998**, *32*, 2429-2441.

(14) Klotz, B.; Sorensen, S.; Barnes, I.; Becker, K. H.; Etzkorn, T.; Volkamer, R.; Platt, U.; Wirtz, K.; Martin-Reviejo, M. Atmospheric oxidation of toluene in a large-volume outdoor photoreactor: In situ determination of ring-retaining product yields *J. Phys. Chem. A* **1998**, *102*, 10289-10299.

(15) Olariu, R. I.; Bejan, I.; Barnes, I.; Klotz, B.; Becker, K. H.; Wirtz, K. Rate coefficients for the gas-phase reaction of NO_3 radicals with selected dihydroxybenzenes *International Journal of Chemical Kinetics* **2004**, *36*, 577-583.

(16) Barnes, I.; Becker, K.H.; Mihalopoulos, N. An FTIR product study of the photooxidation of dimethyl disulfide *Journal of Atmospheric Chemistry* **1994**, *18*, 267-289.

(17) Volkamer, R.; Platt, U.; Wirtz, K. Primary and Secondary Glyoxal Formation from Aromatics: Experimental Evidence for the Bicycloalkyl- Radical Pathway from Benzene, Toluene, and p-Xylene *The Journal of Physical Chemistry A* **2001**, *105*, 7865-7874.

(18) Klotz, B.; Barnes, I.; Becker, K. H. New results on the atmospheric photooxidation of simple alkylbenzenes *Chemical Physics* **1998**, *231*, 289-301.

(19) Olariu, R. I.; Klotz, B.; Barnes, I.; Becker, K. H.; Mocanu, R. FT–IR study of the ring-retaining products from the reaction of OH radicals with phenol, o-, m-, and p-cresol *Atmospheric Environment* **2002**, *36*, 3685-3697.

(20) Klotz, B.; Barnes, I.; Becker, K. H.; Golding, B. T. Atmospheric chemistry of benzene oxide/oxepin *Journal of the Chemical Society, Faraday Transactions* **1997**, *93*, 1507-1516.

(21) Bohn, B. Formation of Peroxy Radicals from OH-Toluene Adducts and O_2 *The Journal of Physical Chemistry A* **2001**, *105*, 6092-6101.

(22) Bohn, B.; Zetzsch, C. Gas-phase reaction of the OH–benzene adduct with O_2: reversibility and secondary formation of HO_2 *Physical Chemistry Chemical Physics* **1999**, *1*, 5097-5107.

(23) Yu, J.; Jeffries, H. E. Atmospheric photooxidation of alkylbenzenes—II. Evidence of formation of epoxide intermediates *Atmospheric Environment* **1997**, *31*, 2281-2287.

6 Indexes

(24) Kwok, E. S. C.; Aschmann, S. M.; Atkinson, R.; Arey, J. Products of the gas-phase reactions of o-, m-and p-xylene with the OH radical in the presence and absence of NOx *Journal of the Chemical Society, Faraday Transactions* **1997**, *93*, 2847-2854.

(25) Thomson, J. J. Bakerian Lecture: Rays of Positive Electricity *Proceedings of the Royal Society of London. Series A* **1913**, *89*, 1-20.

(26) Aebersold, R.; Mann, M. Mass spectrometry-based proteomics *Nature* **2003**, *422*, 198-207.

(27) Noble, C. A.; Prather, K. A. Real-time single particle mass spectrometry: A historical review of a quarter century of the chemical analysis of aerosols *Mass Spectrometry Reviews* **2000**, *19*, 248-274.

(28) Nibbering, N. M. Four decades of joy in mass spectrometry *Mass Spectrometry Reviews* **2006**, *25*, 962-1017.

(29) Gross, M. L.; Caprioli, R. M. In *The Encyclopedia of Mass Spectrometry*; 1 ed.; Caprioli, R. M., Ed.; Elsevier: Oxford, **2007**; Vol. 6.

(30) Van Berkel, G. J. An overview of some recent developments in ionization methods for mass spectrometry *Eur J Mass Spectrom (Chichester, Eng)* **2003**, *9*, 539.

(31) Brunnée, C. New instrumentation in mass spectrometry *International Journal of Mass Spectrometry and Ion Physics* **1982**, *45*, 51-86.

(32) Cooks, R. G.; Hoke Ii, S. H.; Morand, K. L.; Lammert, S. A. Mass spectrometers: instrumentation *International Journal of Mass Spectrometry and Ion Processes* **1992**, *118-119*, 1-36.

(33) Munson, M. S. B.; Field, F. H. Chemical Ionization Mass Spectrometry. I. General Introduction *Journal of the American Chemical Society* **1966**, *88*, 2621-2630.

(34) Niessen, W. M. A. Advances in instrumentation in liquid chromatography-mass spectrometry and related liquid-introduction techniques *Journal of Chromatography A* **1998**, *794*, 407-435.

(35) Bruins, A. P. Atmospheric-pressure-ionization mass spectrometry : I. Instrumentation and ionization techniques *TrAC Trends in Analytical Chemistry* **1994**, *13*, 37-43.

(36) Syage, J. A.; Hanolda, K. A.; Lynna, T. C.; Hornerb, J. A.; Thakur, R. A. Atmospheric pressure photoionization II. Dual source ionization *J. Chromatogr. A* **2004**, *1050*, 137–149.

(37) Constapel, M.; Schellenträger, M.; Schmitz, O. J.; Gäb, S.; Brockmann, K. J.; Giese, R.; Th, B. Atmospheric-pressure laser ionization: a novel ionization method for liquid chromatography/mass spectrometry *Rapid Communications in Mass Spectrometry* **2005**, *19*, 326-336.

6 Indexes

(38) Bruker Daltoniks GmbH *APLI source* http://www.bdal.com/products/lc-ms/aplisource/overview.html **2010**.

(39) Lorenz, M.; Short, L. C.; Constapel, M.; Schellenträger, M.; Möschter, S.; Schmitz, O. J.; Gäb, S.; Brockmann, K.-J.; Droste, S.; Benter, T., ; Atmospheric Pressure Laser Ionization (APLI) Mass Spectrometry for Measurement of Aromatic Hydrocarbons with Low Polarity. *Proceedings of the 53rd ASMS Conference on Mass Spectrometry and Allied Topics;* San Antonio, TX, USA, June **2005**.

(40) Schiewek, R.; Schellenträger, M.; Mönnikes, R.; Lorenz, M.; Giese, R.; Brockmann, K. J.; Gäb, S.; Benter, T.; Schmitz, O. J. Ultrasensitive Determination of Polycyclic Aromatic Compounds with Atmospheric-Pressure Laser Ionization as an Interface for GC/MS *Analytical Chemistry* **2007**, *79*, 4135-4140.

(41) Schiewek, R.; Lorenz, M.; Giese, R.; Brockmann, K.; Benter, T.; Gäb, S.; Schmitz, O. Development of a multipurpose ion source for LC-MS and GC-API MS *Analytical and Bioanalytical Chemistry* **2008**, *392*, 87-96.

(42) Schiewek, R.; Mönnikes, R.; Wulf, V.; Gäb, S.; Brockmann, K.; Benter, T.; Schmitz, O. A Universal Ionization Label for the APLI-(TOF)MS Analysis of Small Molecules and Polymers *Angewandte Chemie International Edition* **2008**, *47*, 9989-9992.

(43) Lorenz, M.; Schiewek, R.; Brockmann, K. J.; Schmitz, O. J.; Gäb, S.; Benter, T. The Distribution of Ion Acceptance in Atmospheric Pressure Ion Sources: Spatially Resolved APLI Measurements *Journal of the American Society for Mass Spectrometry* **2008**, *19*, 400-410.

(44) Kersten, H.; Lorenz, M.; Brockmann, K. J.; Benter, T. Evaluation of the performance of small diode pumped UV solid state (DPSS) Nd:YAG lasers as new radiation sources for Atmospheric Pressure Laser Ionization Mass Spectrometry (APLI-MS) *for publication* **2010**.

(45) Lorenz, M.; Klee, S.; Mönnikes, R.; Mangas Suarez, A. L.; Brockmann, K. J.; Schmitz, O. J.; Gäb, S.; Benter, T. Atmospheric Pressure Laser Ionization (APLI): Investigations on Ion Transport in Atmospheric Pressure Ion Sources. *Proceedings of the 56th ASMS Conference on Mass Spectrometry and Allied Topics;* Denver, CO, USA, June **2008**.

(46) Lorenz, M.; Wißdorf, W.; Klee, S.; Kersten, H.; Brockmann, K. J.; Benter, T. Spatially and temporally resolved atmospheric pressure laser ionization as a powerful tool for the characterization of ion sources: An overview. *Proceedings of the 58th ASMS Conference on Mass Spectrometry and Allied Topics;* Salt Lake City, UT, USA, May **2010**.

(47) Lorenz, M.; Wissdorf, W.; Klee, S.; Kersten, H.; Brockmann, K. J.; Benter, T. Ion Transport Processes in API Sources: Temporally and Spatially Resolved APLI Measurements. *Proceedings of the 57th ASMS Conference on Mass Spectrometry and Allied Topics;* Philadelphia, PA, USA, June **2009**.

6 Indexes

(48) Barnes, I.; Kersten, H.; Wissdorf, W.; Pöhler, T.; Hönen, H.; Klee, S.; Brockmann, K. J.; Benter, T. Novel Laminar Flow Ion Sources for LC- and GC-API MS. *Proceedings of the 58th ASMS Conference on Mass Spectrometry and Allied Topics;* Salt Lake City, UT, USA, May **2010**.

(49) Robb, D. B.; Covey, T. R.; Bruins, A. P. Atmospheric Pressure Photoionization: An Ionization Method for Liquid Chromatography Mass Spectrometry *Analytical Chemistry* **2000**, *72*, 3653-3659.

(50) Syage, J. A.; Evans, M. D.; Hanold, K. A. *Am. Lab.* **2000**, *32*.

(51) Marchi, I.; Rudaz, S.; Veuthey, J.-L. Atmospheric pressure photoionization for coupling liquid-chromatography to mass spectrometry: A review *Talanta* **2009**, *78*, 1-18.

(52) Cheng, B.-M.; Bahou, M.; Chen, W.-C.; Yui, C.-h.; Lee, Y.-P.; Lee, L. C. Experimental and theoretical studies on vacuum ultraviolet absorption cross sections and photodissociation of CH_3OH, CH_3OD, CD_3OH, and CD_3OD *The Journal of Chemical Physics* **2002**, *117*, 1633-1640.

(53) Kanda, K.; Nagata, T.; Ibuki, T. Photodissociation of some simple nitriles in the extreme vacuum ultraviolet region *Chemical Physics* **1999**, *243*, 89-96.

(54) Hatano, Y. Interaction of photons with molecules – cross-sections for photoabsorption, photoionization, and photodissociation *Radiation and Environmental Biophysics* **1999**, *38*, 239-247.

(55) Nee, J. B.; Suto, M.; Lee, L. C. Photoexcitation processes of CH_3OH: Rydberg states and photofragment fluorescence *Chemical Physics* **1985**, *98*, 147-155.

(56) Nuth, J. A.; Glicker, S. The vacuum ultraviolet spectra of HCN, C_2N_2, and CH_3CN *Journal of Quantitative Spectroscopy and Radiative Transfer* **1982**, *28*, 223-231.

(57) Russell, B. R.; Edwards, L. O.; Raymonda, J. W. Vacuum ultraviolet absorption spectra of the chloromethanes *Journal of the American Chemical Society* **1973**, *95*, 2129-2133.

(58) Watanabe, K.; Zelikoff, M.; Edward, C. Y. Absorption coefficients of several atmospheric gases *AFCRC Technical Report No. 53-23* **1953**.

(59) Voinov, V. G.; Vasil'ev, Y. V.; Ji, H.; Figard, B.; Morre`, J.; Egan, T. F.; Barofsky, D. F.; Deinzer, M. L. A Gas Chromatograph/Resonant Electron Capture-TOF Mass Spectrometer for Four Dimensions of Negative Ion Analytical Information *Analytical Chemistry* **2004**, *76*, 2951-2957.

(60) Williamson, D. H.; Knighton, W. B.; Grimsrud, E. P. Effect of buffer gas alterations on the thermal electron attachment and detachment reactions of azulene by pulsed high pressure mass spectrometry *International Journal of Mass Spectrometry* **2000**, *195-196*, 481-489.

(61) Kersten, H.; Funcke, V.; Lorenz, M.; Brockmann, K. J.; Benter, T.; O'Brien, R. Evidence of Neutral Radical Induced Analyte Ion Transformations in APPI and Near-VUV APLI *Journal of the American Society for Mass Spectrometry* **2009**, *20*, 1868-1880.

(62) Gioumousis, G.; Stevenson, D. P. Reactions of Gaseous Molecule Ions with Gaseous Molecules .5. Theory *Journal of Chemical Physics* **1958**, *29*, 294-299.

(63) Eichelberger, B. R.; Snow, T. P.; Bierbaum, V. M. Collision rate constants for polarizable ions *American Society for Mass Spectrometry* **2003**, *14*, 501-505.

(64) Syage, J. A. Mechanism of [M+H]$^+$ Formation in Photoionization Mass Spectrometry *American Society for Mass Spectrometry* **2004**, *15*, 1521-1533.

(65) Short, L. C.; Cai, S. S.; Syage, J. A. APPI-MS: Effects of mobile phases and VUV lamps on the detection of PAH compounds *Journal of the American Society for Mass Spectrometry* **2007**, *18*, 589-599.

(66) Robb, D. B.; Blades, M. W. Factors affecting primary ionization in dopant-assisted atmospheric pressure photoionization (DA-APPI) for LC/MS *Journal of the American Society for Mass Spectrometry* **2006**, *17*, 130-138.

(67) Robb, D. B.; Blades, M. W. Effects of solvent flow, dopant flow, and lamp current on dopant-assisted atmospheric pressure photoionization (DA-APPI) for LC-MS. Ionization via proton transfer *Journal of the American Society for Mass Spectrometry* **2005**, *16*, 1275-1290.

(68) Orlandini, I.; Riedel, U. Chemical kinetics of NO removal by pulsed corona discharges *Journal of Physics D: Applied Physics* **2000**, *33*, 2467.

(69) Anicich An Index of the Literature for Bimolecular Gas Phase Cation-Molecule Reaction Kinetics *JPL Publication 03-19* **2003**.

(70) Le Page, V.; Keheyanb, Y.; Snow, T. P.; Bierbaum, V. M. Gas phase chemistry of pyrene and related cations with molecules and atoms of interstellar interest *International Journal of Mass Spectrometry* **1999**, *187*, 949–959.

(71) Mahle, N. H.; Cooks, R. G.; Korzeniowski, R. W. Hydroxylation of Aromatic-Hydrocarbons in Mass-Spectrometer Ion Sources under Atmospheric-Pressure Ionization and Chemical Ionization Conditions *Analytical Chemistry* **1983**, *55*, 2272-2275.

(72) Brian L. Frey, Y. L., Michael S. Westphall, Lloyd M. Smith Controlling Gas-Phase Reactions for Efficient Charge Reduction Electrospray Mass Spectrometry of Intact Proteins *Journal of the American Society for Mass Spectrometry* **2005**, *16*, 1876-1887.

(73) Kauppila, T. J.; Kuuranne, T.; Meurer, E. C.; Eberlin, M. N.; Kotiaho, T.; Kostiainen, R. Atmospheric Pressure Photoionization Mass Spectrometry. Ionization Mechanism and the Effect of Solvent on the Ionization of Naphthalenes *Analytical Chemistry* **2002**, *74*, 5470-5479.

6 Indexes

(74) Bruker Daltoniks GmbH HCTultra User Manual; Version 1.2.1 **2006**.

(75) March, R.; Hughes, J. *Quadrupole storage mass spectrometry*; John Wiley & Sons, Inc.: New York, NY, USA **1989**.

(76) March, R. E. An Introduction to Quadrupole Ion Trap Mass Spectrometry *Journal of Mass Spectrometry* **1997**, *32*, 351-369.

(77) Guan, S.; Marshall, A. G. Equilibrium space charge distribution in a quadrupole ion trap *Journal of the American Society for Mass Spectrometry* **1994**, *5*, 64-71.

(78) Schiewek, R. *Entwicklung einer Multi-Purpose Ionenquelle für die AP-MS sowie Design und Anwendung von APLI-Ionisationslabeln* Dissertation, Bergische Universität Wuppertal, **2008**; urn:nbn:de:hbz:468-20090039.

(79) Patent application *Atmosphärendruck-Ionenquelle hoher Ausbeute für Vakuum-Ionenspektrometer*; DE 10 2009 037 716.6; Germany; **2009**.

(80) Patent application *High Yield Atmospheric Pressure Ion Source for Ion Spectrometers in Vacuum*; 1013684.4; United Kingdom; **2010**.

(81) Kersten, H.; Wissdorf, W.; Brockmann, K. J.; Benter, T.; O'Brien, R. VUV Photoionization within Transfer Capillaries of Atmospheric Pressure Ion sources. *Proceedings of the 58th ASMS Conference on Mass Spectrometry and Allied Topics;* Salt Lake City, UT, USA, **2010**.

(82) Kersten, H.; Funcke, V.; Lorenz, M.; Brockmann, K. J.; Benter, T.; O'Brien, R. Evidence of Neutral Radical Induced Analyte Ion Transformations in APPI and Near-VUV APLI *Journal of the American Society for Mass Spectrometry* **2009**, *20*, 1868-1880.

(83) Ruzicka, K.; Mokbel, I.; Majer, V.; Ruzicka, V.; Jose, J.; Zabransky, M. Description of vapour-liquid and vapour-solid equilibria for a group of polycondensed compounds of petroleum interest *Fluid Phase Equilibria* **1998**, *148*, 107-137.

(84) *Organic Synthesis*; Blatt, A. H., Ed.; Wiley: New York, NY, USA, 1943; Vol. 2.

(85) Jelezniak, M.; Jelezniak, I. *CHEMKED: Chemical Kinetics of Gas Phase Reactions Version 3.3*; Darmstadt, Germany, **2007**.

(86) M. J. Frisch, G. W. T., H. B. Schlegel, G. E. Scuseria, M. A. Robb, J. R. Cheeseman, J. A. Montgomery, Jr., T. Vreven, K. N. Kudin, J. C. Burant, J. M. Millam, S. S. Iyengar, J. Tomasi, V. Barone, B. Mennucci, M. Cossi, G. Scalmani, N. Rega, G. A. Petersson, H. Nakatsuji, M. Hada, M. Ehara, K. Toyota, R. Fukuda, J. Hasegawa, M. Ishida, T. Nakajima, Y. Honda, O. Kitao, H. Nakai, M. Klene, X. Li, J. E. Knox, H. P. Hratchian, J. B. Cross, C. Adamo, J. Jaramillo, R. Gomperts, R. E. Stratmann, O. Yazyev, A. J. Austin, R. Cammi, C. Pomelli, J. W. Ochterski, P. Y. Ayala, K. Morokuma, G. A. Voth, P. Salvador, J. J. Dannenberg, V. G. Zakrzewski, S. Dapprich, A. D. Daniels, M. C. Strain, O. Farkas, D. K. Malick, A. D. Rabuck, K. Raghavachari, J. B. Foresman, J. V. Ortiz, Q. Cui, A. G. Baboul, S. Clifford, J. Cioslowski, B. B.

Stefanov, G. Liu, A. Liashenko, P. Piskorz, I. Komaromi, R. L. Martin, D. J. Fox, T. Keith, M. A. Al-Laham, C. Y. Peng, A. Nanayakkara, M. Challacombe, P. M. W. Gill, B. Johnson, W. Chen, M. W. Wong, C. Gonzalez, and J. A. Pople *Gaussian 03 Revision-D.01 Gaussian 03 Revision-D.01*; Gaussian, Inc.; Pittsburgh PA, **2003**.

(87) Dennington, R.; Keith, T.; Millam, J. *GaussView Version 4.1*; Semichem, Inc., Shawnee Mission, KS, **2007**.

(88) Stephens, P. J.; Devlin, F. J.; Chabalowski, C. F.; Frisch, M. J. Ab-Initio Calculation of Vibrational Absorption and Circular-Dichroism Spectra Using Density-Functional Force-Fields *Journal of Physical Chemistry* **1994**, *98*, 11623-11627.

(89) Garifzianova, G. G.; Tsyshevskii, R. V.; Shamov, A. G.; Khrapkovskii, G. M. A quantum-chemical study of n-butane and of butane cation radical *International Journal of Quantum Chemistry* **2007**, *107*, 2489-2493.

(90) Laali, K. K.; Okazaki, T.; Galembeck, S. E. Stable ion and electrophilic chemistry of fluoranthene-PAHs *Journal of the Chemical Society-Perkin Transactions 2* **2002**, 621-629.

(91) Langhoff, S. R. Theoretical infrared spectra for polycyclic aromatic hydrocarbon neutrals, cations, and anions *Journal of Physical Chemistry* **1996**, *100*, 2819-2841.

(92) Rodriquez, C. F.; Shoeib, T.; Chu, I. K.; Siu, K. W. M.; Hopkinson, A. C. Comparison between protonation, lithiation, and argentination of 5-oxazolones: A study of a key intermediate in gas-phase peptide sequencing *Journal of Physical Chemistry A* **2000**, *104*, 5335-5342.

(93) Manura, D. J.; Dahl, D. A. *SIMION 8.0*; Scientific Instrument Services, Inc; **2007**.

(94) Thermo Scientific *Ion Max Source* https://thermoscientific.com/wps/portal/ts/products/ detail?navigationId=L10963&cate goryId=87462&productId=11962148 **2010**.

(95) Waters Xevo API sources http://www.waters.com/waters/nav.htm?locale=de_DE&cid=10156110# **2010**.

(96) Wißdorf, W.; Pohler, L.; Pöhler, T.; Hönen, H.; Brockmann, K. J.; Benter, T. Modular computational toolset for atmospheric pressure ionization method development: SIMION meets FEM. *Proceedings of the 58th ASMS Conference on Mass Spectrometry and Allied Topics* Salt Lake City, UT, USA, May **2010**.

(97) Wissdorf, W.; Brockmann, K.J.; Benter, T *in preperation*.

(98) Pöhler, T.; Hoenen, H; Wissdorf, W., Brockmann, K.J., Benter, T. *in preperation*.

(99) Wutz, M.; Adam, H.; Walcher, W., Eds. *Theorie und Praxis der Vakuumtechnik*; 4th ed.; Friedr. Vieweg und Sohn: Braunschweig/Wiesbaden, Germany, **1988**.

(100) Gustafson, K. E.; Dickhut, R. M. Molecular Diffusivity of Polycyclic Aromatic-Hydrocarbons in Air *Journal of Chemical and Engineering Data* **1994**, *39*, 286-289.

6 Indexes

(101) ATL Lasertechnik GmbH *Users Manual Excimer Laser ATLEX-SI* **2008**.

(102) Zakheim, D. S.; Johnson, P. M. Rate equation modelling of molecular multiphoton ionization dynamics *Chemical Physics* **1980**, *46*, 263-272.

(103) Novotny, O.; Sivaraman, B.; Rebrion-Rowe, C.; Travers, D.; Biennier, L.; Mitchell, J. B. A.; Rowe, B. R. Recombination of polycyclic aromatic hydrocarbon photoions with electrons in a flowing afterglow plasma *Journal of Chemical Physics* **2005**, *123*, 104303-1 - 104303-6

(104) Baker, B. G.; Johnson, B. B.; Maire, G. L. C. Photoelectric work function measurements on nickel crystals and films *Surface Science* **1971**, *24*, 572-586.

(105) Logothetis, E. M.; Hartman, P. L. Laser-Induced Electron Emission from Solids: Many-Photon Photoelectric Effects and Thermionic Emission *Physical Review* **1969**, *187*, 460.

(106) Brockmann, K. J.; Wissdorf, W.; Hyzak, L.; Kersten, H.; Benter, T. Fundamental characterization of ion transfer capillaries used in Atmospheric Pressure Ionization sources. *Proceedings of the 58th ASMS Conference on Mass Spectrometry and Allied Topics;* Salt Lake City, UT, USA, **2010**.

(107) Chen, C. K.; de Castro, A. R. B.; Shen, Y. R. Surface-Enhanced Second-Harmonic Generation *Physical Review Letters* **1981**, *46*, 145.

(108) Tomas, C.; Vinet, E.; Girardeau-Montaut, J. P. Simultaneous measurements of second-harmonic generation and two-photon photoelectric emission from Au *Applied Physics A: Materials Science & Processing* **1999**, *68*, 315-320.

(109) Lin, B. W.; Sunner, J. Ion-Transport by Viscous-Gas Flow-through Capillaries *Journal of the American Society for Mass Spectrometry* **1994**, *5*, 873-885.

(110) Michalke, A. Beitrag zur Rohrströmung kompressibler Fluide mit Reibung und Wärmeübergang *Archive of Applied Mechanics* **1987**, *57*, 377-392.

(111) Hanna, S. J.; Campuzano-Jost, P.; Simpson, E. A.; Robb, D. B.; Burak, I.; Blades, M. W.; Hepburn, J. W.; Bertram, A. K. A new broadly tunable (7.4-10.2 eV) laser based VUV light source and its first application to aerosol mass spectrometry *International Journal of Mass Spectrometry* **2009**, *279*, 134-146.

(112) Mühlberger, F.; Wieser, J.; Ulrich, A.; Zimmermann, R. Single photon ionization (SPI) via incoherent VUV-excimer light: Robust and compact time-of-flight mass spectrometer for on-line, real-time process gas analysis *Analytical Chemistry* **2002**, *74*, 3790-3801.

(113) Efthimiopoulos, T.; Zouridis, D.; Ulrich, A. Excimer emission spectra of rare gas mixtures using either a supersonic expansion or a heavy-ion-beam excitation *Journal of Physics D: Applied Physics* **1997**, *30*, 1746-1754.

6 Indexes

(114) Millet, P.; Birot, A.; Brunet, H.; Dijolis, H.; Galy, J.; Salamero, Y. Spectroscopic and kinetic analysis of the VUV emissions of argon and argon-xenon mixtures. I. Study of pure argon *Journal of Physics B: Atomic and Molecular Physics* **1982**, *15*, 2935-2944.

(115) Wieser, J.; Murnick, D. E.; Ulrich, A.; Huggins, H. A.; Liddle, A.; Brown, W. L. Vacuum ultraviolet rare gas excimer light source *Review of Scientific Instruments* **1997**, *68*, 1360-1364.

(116) Jensen, C. A.; Libby, W. F. Intense 584-Å Light from a Simple Continuous Helium Plasma *Physical Review* **1964**, *135*, 1247-1252.

(117) Moselhy, M.; Schoenbach, K. H. Excimer emission from cathode boundary layer discharges *Journal of Applied Physics* **2004**, *95*.

(118) Gellert, B.; Kogelschatz, U. Generation of excimer emission in dielectric barrier discharges *Applied Physics B: Lasers and Optics* **1991**, *52*, 14-21.

(119) Kogelschatz, U. Silent discharges for the generation of ultraviolet and vacuum ultraviolet excimer radiation *Pure & Applied Chemistry* **1990**, *62*, 1667–1674.

(120) Sankaran, R. M.; Giapis, K. P.; Moselhy, M.; Schoenbach, K. H. Argon excimer emission from high-pressure microdischarges in metal capillaries *Applied Physics Letters* **2003**, *83*, 4728-4730.

(121) Laqua, K. In *Ullmanns Enczklopädie der technischen Chemie*; 4 ed.; Kelker, H., Ed.; Verlag Chemie: Weinheim, **1980**; Vol. 5, p 441-500.

(122) Paschen, F. Ueber die zum Funkenübergang in Luft, Wasserstoff und Kohlensäure bei verschiedenen Drucken erforderliche Potentialdifferenz *Annalen der Physik* **1889**, *273*, 69-96.

(123) Loeb, L. B. The Problem of the Mechanism of Static Spark Discharge *Reviews of Modern Physics* **1936**, *8*, 267.

(124) Druyvesteyn, M. J.; Penning, F. M. The Mechanism of Electrical Discharges in Gases of Low Pressure *Reviews of Modern Physics* **1940**, *12*, 87.

(125) Kersten, H.; Carter, D.; Brockmann, K. J.; Benter, T.; O'Brien, R. R. Insights gained using Atmospheric Pressure Electrical Discharge as a photon source for Atmospheric Pressure Photoionization Mass Spectrometry. *The 18th International Mass Spectrometry Conference;* Bremen, Germany, August **2009**.

(126) HartlauerPräzisionsElektronikGmbH *Spezifikation, Einbau-und Gebrauchsanleitung für HV-Netzteil DD20_10 Sonderausführung C-Lader* Grassau, Germany, **2010** .

(127) Knop, H.; Uhrig, M.; Berkemeier, M.; Becker, K.; Hanne, G. F. A temperature-stabilized LiF line filter for the argon 106.7 nm resonance line *Measurement Science and Technology* **1997**, *8*.

6 Indexes

(128) Inn, E. C. Y. Vacuum ultraviolet spectroscopy: A review *Spectrochimica Acta* **1955**, *7*, 65-87.

(129) Duncanson, A.; Stevenson, R. W. H. Some Properties of Magnesium Fluoride crystallized from the Melt *Proceedings of the Physical Society* **1958**, *72*, 1001.

(130) Korth Kristalle GmbH *VUV-Transmissionsspektrum von LiF* http://www.korth.de/de/503728952d091450d/503728952d0b33731.htm **2010**.

(131) Kaiser, H.; Specker, H. Bewertung und Vergleich von Analysenverfahren *Fresenius' Journal of Analytical Chemistry* **1956**, *149*, 46-66.

(132) Ledernez, L.; Olcaytug, F.; Yusada, H.; Urban, G. A modification of Paschen law for Argon *ICPIG;* Cancun, Mexico, **2009**.

(133) McDaniel, E. W., Ed *Collision Phenomena in Ionized Gases*; John Wiley & Sons, Inc.: New York, NY, USA, **1964**.

(134) Brode, R. B. The Quantitative Study of the Collisions of Electrons with Atoms *Reviews of Modern Physics* **1933**, *5*, 257.

(135) Ralchenko, Y.; Kramida, A. E.; Reader, J.; NISTASDTeam(2008) *NIST Atomic Spectra Database (version 3.1.5)* http://physics.nist.gov/asd3 **2008**.

(136) Koehler, H. A.; Ferderber, L. J.; Redhead, D. L.; Ebert, P. J. Vacuum-ultraviolet emission from high-pressure xenon and argon excited by high-current relativistic electron beams *Physical Review A* **1974**, *9*, 768.

(137) Moselhy, M.; Stark, R. H.; Schoenbach, K. H.; Kogelschatz, U. Resonant energy transfer from argon dimers to atomic oxygen in microhollow cathode discharges *Applied Physics Letters* **2001**, *78*, 880-882.

(138) Manion, J. A.; Huie, R. E.; Levin, R. D.; Burgess, D. R.; Orkin, V. L.; Tsang, W.; McGivern, W. S.; Hudgens, J. W.; Knayzev, V. D.; Atkinson, D. B.; Chai, E.; Tereza, A. M.; Lin, C.-Y.; Allison, T. C.; Mallard, W. G.; Westley, F.; Herron, J. T.; Hampson, R. F.; Frizzel, D. H. *NIST Chemical Kinetics Database, NIST Standard Reference Database 17, version 7.0 (Web Version), Release 1.5,* http://kinetics.nist.gov/ **2010**.

(139) Gingell, M.; et al. The electronic states of cyclopropane studied by VUV absorption and electron energy-loss spectroscopies *Journal of Physics B: Atomic, Molecular and Optical Physics* **1999**, *32*, 2729.

(140) Wood, J. M.; Kahr, B.; Hoke, S. H.; Dejarme, L.; Cooks, R. G.; Benamotz, D. Oxygen and Methylene Adducts of C-60 and C-70 *Journal of the American Chemical Society* **1991**, *113*, 5907-5908.

(141) Lifshitz, C. Energetics and dynamics through time-resolved measurements in mass spectrometry: Aromatic hydrocarbons, polycyclic aromatic hydrocarbons and fullerenes *International Reviews in Physical Chemistry* **1997**, *16*, 113-139.

(142) Ling, Y.; Gotkis, Y.; Lifshitz, C. Time-Dependent Mass-Spectra and Breakdown Graphs .18. Pyrene *European Mass Spectrometry* **1995**, *1*, 41-49.

(143) Lias, S. G.; Bartmess, J. E.; Liebmann, J. F.; Holmes, J. L.; Levin, R. D.; Mallard, W. G. *Gas-phase ion and neutral thermochemistry*; American Institute of Physics: Melville, NY, USA, **1988**; Vol. 17.

(144) Herzberg, G. *Molecular spectra and molecular structure*; Van Nostrand Reinhold Company: New York, NY, USA, **1966**.

(145) Nourse, B. D.; Cox, K. A.; Cooks, R. G. Ion-Molecule Chemistry of Pyrene in an Ion-Trap Mass-Spectrometer *Organic Mass Spectrometry* **1992**, *27*, 453-462.

(146) Jones, C. M.; Asher, S. A. Ultraviolet resonance Raman study of the pyrene S_4, S_3, and S_2 excited electronic states *The Journal of Chemical Physics* **1988**, *89*, 2649-2661.

(147) Baba, H.; Aoi, M. Vapor-Phase Fluorescence-Spectra from Second Excited Singlet-State of Pyrene and Its Derivatives *Journal of Molecular Spectroscopy* **1973**, *46*, 214-222.

(148) Baba, H.; Nakajima, A.; Aoi, M.; Chihara, K. Fluorescence from the Second Excited Singlet State and Radiationless Processes in Pyrene Vapor *The Journal of Chemical Physics* **1971**, *55*, 2433-2438.

(149) Borisevich, N. A.; Vodovatov, L. B.; Dyachenko, G. G.; Petukhov, V. A.; Semenov, M. A. Spectroscopy of pyrene clusters formed in a supersonic jet *Laser Physics* **1997**, *7*, 400-402.

(150) Chihara, K.; Baba, H. Effects of Foreign Gases on Dual Fluorescences of Pyrene Vapor *Bulletin of the Chemical Society of Japan* **1975**, *48*, 3093-3100.

(151) Chihara, K.; Baba, H. Quenching of Dual Fluorescences of Pyrene Vapor by High-Pressure Oxygen or Nitric-Oxide *Chemical Physics* **1977**, *25*, 299-306.

(152) Mendes, M. A.; Moraes, L. A. B.; Sparrapan, R.; Eberlin, M. N.; Kostiainen, R.; Kotiaho, T. Oxygen atom transfer to positive ions: A novel reaction of ozone in the gas phase *Journal of the American Chemical Society* **1998**, *120*, 7869-7874.

(153) Atkinson, R.; Arey, J.; Zielinska, B.; Aschmann, S. M. Kinetics and Nitro-Products of the Gas-Phase Oh and No3 Radical-Initiated Reactions of Naphthalene-D8, Fluoranthene-D10, and Pyrene *International Journal of Chemical Kinetics* **1990**, *22*, 999-1014.

(154) Frerichs, H.; Tappe, M.; Wagner, H. G. Comparison of the Reactions of Monocyclic and Polycyclic Aromatic-Hydrocarbons with Oxygen-Atoms *Berichte Der Bunsen-Gesellschaft-Physical Chemistry Chemical Physics* **1990**, *94*, 1404-1407.

(155) Page, V. L. Chemical constraints on organic cations in the interstellar medium *J. Am. Chem. Soc.* **1997**, *119*, 8373-8374.

6 Indexes

(156) MacNeil, K. A. G.; Futrell, J. H. Ion-molecule reactions in gaseous acetone **1972**, *76*, 409-415.

(157) Shimamori, H.; Fessenden, R. W. Thermal Electron-Attachment to Oxygen and Vanderwaals Molecules Containing Oxygen *Journal of Chemical Physics* **1981**, *74*, 453-466.

(158) Mock, R. S.; Grimsrud, E. P. Gas-phase electron photodetachment spectroscopy of the molecular anions of nitroaromatic hydrocarbons at atmospheric pressure *Journal of the American Chemical Society* **1989**, *111*, 2861-2870.

(159) Shelley, J. T.; Wiley, J. S.; Chan, G. C. Y.; Schilling, G. D.; Ray, S. J.; Hieftje, G. M. Characterization of Direct-Current Atmospheric-Pressure Discharges Useful for Ambient Desorption/Ionization Mass Spectrometry *Journal of the American Society for Mass Spectrometry* **2009**, *20*, 837-844.

(160) Syage, J. A.; Hanold, K. A.; Lynn, T. C.; Horner, J. A.; Thakur, R. A. Atmospheric pressure photoionization II. Dual source ionization *Journal of Chromatography A* **2004**, *1050*, 137-149.

Die VDM Verlagsservicegesellschaft sucht für wissenschaftliche Verlage abgeschlossene und herausragende

Dissertationen, Habilitationen, Diplomarbeiten, Master Theses, Magisterarbeiten usw.

für die kostenlose Publikation als Fachbuch.

Sie verfügen über eine Arbeit, die hohen inhaltlichen und formalen Ansprüchen genügt, und haben Interesse an einer honorarvergüteten Publikation?

Dann senden Sie bitte erste Informationen über sich und Ihre Arbeit per Email an *info@vdm-vsg.de*.

Sie erhalten kurzfristig unser Feedback!

VDM Verlagsservicegesellschaft mbH
Dudweiler Landstr. 99
D - 66123 Saarbrücken

Telefon +49 681 3720 174
Fax +49 681 3720 1749

www.vdm-vsg.de

Die VDM Verlagsservicegesellschaft mbH vertritt

Printed by Books on Demand GmbH, Norderstedt / Germany